Platinum Group Metals and Compounds

A symposium sponsored by
the Division of Inorganic
Chemistry at the 158th
Meeting of the American
Chemical Society,
New York, N. Y.,
Sept. 8–9, 1969.

U. V. Rao

Symposium Chairman

ADVANCES IN CHEMISTRY SERIES **98**

AMERICAN CHEMICAL SOCIETY

WASHINGTON, D. C.   1971

Coden: ADCSHA

Copyright © 1971

American Chemical Society

All Rights Reserved

Library of Congress Catalog Card 76-152,755

ISBN 8412-0135-8

PRINTED IN THE UNITED STATES OF AMERICA

Advances in Chemistry Series
Robert F. Gould, *Editor*

Advisory Board

Paul N. Craig

Thomas H. Donnelly

Gunther Eichhorn

Frederick M. Fowkes

Fred W. McLafferty

William E. Parham

Aaron A. Rosen

Charles N. Satterfield

Jack Weiner

AMERICAN CHEMICAL SOCIETY PUBLICATIONS

FOREWORD

ADVANCES IN CHEMISTRY SERIES was founded in 1949 by the American Chemical Society as an outlet for symposia and collections of data in special areas of topical interest that could not be accommodated in the Society's journals. It provides a medium for symposia that would otherwise be fragmented, their papers distributed among several journals or not published at all. Papers are refereed critically according to ACS editorial standards and receive the careful attention and processing characteristic of ACS publications. Papers published in ADVANCES IN CHEMISTRY SERIES are original contributions not published elsewhere in whole or major part and include reports of research as well as reviews since symposia may embrace both types of presentation.

CONTENTS

Preface vii

1. Magnetic Properties of the Platinum Metals and Their Alloys 1
   H. J. Albert and L. R. Rubin

2. Platinum Metal Chalcogenides 17
   Aaron Wold

3. Synthesis and Crystal Chemistry of Some New Complex Palladium Oxides 28
   Olaf Muller and Rustum Roy

4. Reaction of Platinum Dioxide with Some Metal Oxides 39
   Henry R. Hoekstra, Stanley Siegel, and Francis X. Gallagher

5. Nitrido Complexes of the Platinum Group Metals 54
   M. J. Cleare and F. M. Lever

6. Hydrido Complexes of Platinum Group Metals 66
   L. M. Venanzi

7. Ultraviolet Spectroscopic Qualities of Platinum Group Compounds 74
   Don S. Martin, Jr.

8. NMR Spectra of Compounds of the Platinum Group Metals 98
   R. Stuart Tobias

9. Crystal Structures of Complexes of the Platinum Group Metals ... 120
   E. L. Amma

10. ^{195}Pt—A Survey of Mossbauer Spectroscopy 135
    N. Benczer-Koller

11. Double Bond Isomerization as a Product-Controlling Factor in Hydrogenation Over Platinum Group Metals 150
    Paul N. Rylander

Index 163

PREFACE

Almost every chemist is aware of the fact that platinum group metal chemistry usually centers around the phenomenon of catalysis. Other interesting phases of the chemistry of the platinum group metals which could be of interest to inorganic chemists are usually obscured by this singular though significant application. When I approached Eugene Rachow, Chairman of the Symposium Planning Committee of the Inorganic Division of ACS, I had every intention of covering all facets of the platinum group metal chemistry. Because of the limitations of time, I was obliged to limit the symposium to four sessions. The material to be presented at the symposium was divided into four categories: (1) Synthesis, Structure, Determination, and Study of Magnetic and Thermodynamic Properties of Platinum Group Metal Compounds, (2) Recent Chemistry of σ- and π-Bonded Complexes of Platinum Group Metals, (3) Spectroscopic Properties of Platinum Group Metal Compounds, (4) Newer Industrial Aspects of Platinum Group Metals and Their Compounds.

Each category, by itself, can be a topic for a separate symposium, and it is hoped that in the future such symposia will be forthcoming. The topics chosen were such that they would cover a broad area of the chemistry of platinum group metals. The session on industrial aspects was included to enable scientists in industry to present their views on the problems that are facing the industry and perhaps stimulate sufficient interest so that newer applications could be developed in the future. It is also hoped that such a forum would enable scientists in industry to summarize broadly the work carried out by them without worry about violation of proprietary nature of the work.

I would like to take this opportunity to extend my thanks to all the speakers at the symposium and to all the authors who contributed articles to this volume. I would also wish to express my gratitude to the management of Matthey Bishop, Inc., who kindly permitted me to utilize their time and facilities to arrange this symposium. My special thanks to G. Cohn of Engelhard Industries and L. B. Hunt of Johnson Matthey & Co. for their cooperation and help rendered to find suitable speakers.

U. V. Rao

Matthey Bishop, Inc.
Malvern, Pa. 19355
December 1970

Magnetic Properties of the Platinum Metals and Their Alloys

H. J. ALBERT and L. R. RUBIN

Engelhard Minerals and Chemicals Corp., Newark, N. J. 07105

> *Although they are paramagnetic, the platinum metals, especially platinum, palladium, and rhodium, are capable of interacting in alloys with other metals to form ferromagnetic or very nearly ferromagnetic materials. Dilute additions of elements of the iron group and its neighbors with platinum and palladium take on enhanced moments which are interpreted as arising from an interaction of the solute moment with the 4d or 5d host electrons. Ferromagnetic and antiferromagnetic structures may result from greater additions of the iron group to the platinum metals. Examples are $FeRh$, Pt_3Fe, Pd_3Fe, and $PtCo$. Compounds of the rare earths with the platinum metals also form magnetic structures with unusual properties.*

The platinum metals—ruthenium, rhodium, and palladium in the 2nd long row of the periodic system and osmium, iridium, and platinum in the 3rd long row—are all paramagnetic. That is, none of these elements have a permanent magnetic moment associated with it. However, as shown in Table I, the mass susceptibility of these elements varies widely over two orders of magnitude. Further, the "nonmagnetic" platinum metals are the elements immediately beneath the "magnetic" series iron–cobalt–nickel and, of course, are all in the transition group. Platinum, palladium, and rhodium form a number of ferromagnetic alloys with the iron group metals and with manganese. The present paper is not an exhaustive review but presents some of the more interesting magnetic work on the platinum metals which has appeared in the last 10–20 years.

Palladium has the highest magnetic susceptibility of the platinum group, and because it is so high it has been termed an "incipient ferromagnet." The implication is that palladium could be induced to become

**Table I. Mass Susceptibility of the Platinum Metals
($\times$ 10^{-6} cgs Units)** [a]

Ru(+0.427)[b]	Rh(+0.9903)[c]	Pd(+5.231)[c]
Os(+0.052)[b]	Ir(+0.133)[b]	Pt(+0.9712)[c]

[a] Reprinted from Engelhard Industries Technical Bulletin.
[b] Determinations made at 25°C.
[c] Determinations made at 20°C.

ferromagnetic. In a weakly paramagnetic material, the susceptibility of the material is independent of temperature. Figure 1 (*39*) shows this to be quite true for rhodium. Palladium, on the other hand, exhibits a strong temperature dependence of susceptibility with a peak at about 70°K. This has been taken in the past as evidence for an antiferromagnetic transition. However, neutron diffraction has shown no evidence of antiferromagnetism, and it now seems likely that this effect is caused by the Fermi surface of palladium being associated with an extremely sharp peak in the density-of-states. The susceptibility *vs.* temperature curve for the palladium–5% rhodium alloy indicates the sensitivity of palladium-based alloys to electron concentration and the density-of-states in relation to the Fermi level (*13*).

Iron impurities greatly complicate the susceptibility measurements on palladium. As noted above, palladium is close to being ferromagnetic,

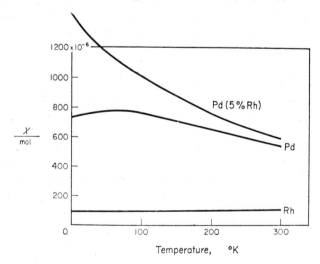

Weiss, J. R., "Solid State Physics for Metallurgists," Pergamon

Figure 1. The susceptibility (10^{-6} ergs/gauss^2/mole) for palladium, rhodium, and palladium–5% rhodium as a function of temperature (*39*)

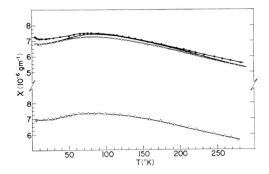

Figure 2. Susceptibility vs. temperature for several "pure" palladium samples (17)

● 3 ppm iron, ○ 2 ppm iron, △ zone refined, solid line less than 1 ppm iron

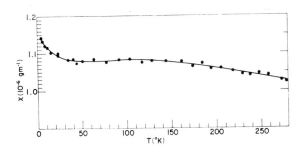

Figure 3. Susceptibility vs. temperature for platinum with approximately 3 ppm of iron (17)

and small additions of iron will form ferromagnetic alloys with Curie temperatures in the cryogenic range. Thus, as shown in Figure 2 (17), even parts per million of iron in palladium may be noticed readily in susceptibility measurements at low temperatures. Neutron scattering measurements have shown that iron impurities have an effect far beyond the nearest-neighbor palladium atoms, extending perhaps to the nearest 100 palladium atoms, accounting for the extreme sensitivity of palladium to iron impurities (28).

Platinum, Figure 3 (17), shows behavior similar to that of palladium but its susceptibility and the effect of iron impurities are much smaller. The small peak at 100°K is analogous to that for palladium.

Local Moments

Considerable research in recent years has been carried out on the magnetic properties of alloys of elements of the second and third long rows of the periodic table with small amounts of elements in the first long row. Some of the most interesting results of this work are centered on the platinum metals. One aspect of the results is given in Figure 4 (*11*), showing the magnetic moment of an iron atom dissolved in the second row transition metals. An enhancement of the iron moment is apparent for alloys with electron concentrations of about 5.5 to 7 and 8.25 to 9. For electron concentrations from 9 to 10.25 there is a very large enhancement, the moment exceeding that of iron in its bulk state (about 2.2 μ_B). Alloys showing this large enhancement are regarded as having "giant moments." Enhancement effects of the solute moment have been found in many alloy systems, including Cr (*2*), Mn (*2*), Fe (*10, 11, 18*), and Co (*9, 10, 38*) in palladium and alloys of palladium and rhodium. The experimental technique widely used to show the existence of such local moments is the measurement of susceptibility. The susceptibility is measured as a function of temperature, usually from 1.4°K to room temperature or higher. The data from alloys with a temperature-dependent susceptibility are fitted to a Curie-Weiss type of curve which leads to

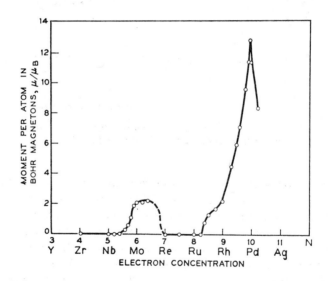

Physical Review

Figure 4. Magnetic moment in Bohr magnetons of an iron atom dissolved in various second row transition metals and alloys (one atomic per cent iron in each metal and alloy) as a function of electron concentration (11)

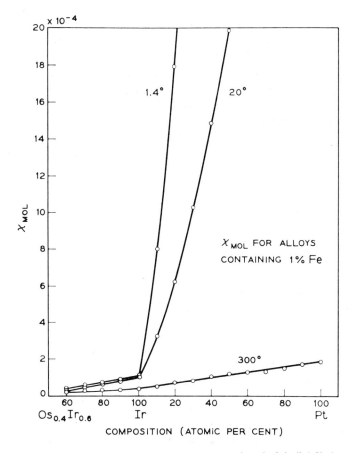

Figure 5. Susceptibility per mole for alloys containing one atomic per cent iron at 3 different temperatures (18)

the interpretation that a local moment exists on the solute atom and permits the calculation of the moment associated with the alloy.

Alloys of iridium and platinum with similar small amounts of iron also exhibit enhancement effects and a giant moment in the case of iron in platinum and platinum-rich alloys as shown in Figure 5 (*18*).

It is obvious that, in view of the large measured moments on the solute atoms, there is more involved than simply the magnetic moment of the solute atom. The postulate most widely used suggests a model in which the moment on the solute atoms interacts with the magnetic moments on the itinerant $4d$ or $5d$ electrons of the host metal. This interaction produces a polarization of the spins in the $4d$ or $5d$ band, at the same time aligning the localized moments on the solute atoms in the same direction as the polarized d-electrons.

Using this local moment model, and using band theory or its variations, a number of workers have been able to formulate expressions which represent the measured magnetic data reasonably well, at least for the case where well-localized moments are developed on the solute atoms (*11, 18*). However, considerably more data has become available on other properties of dilute alloys, including data on resistivity and specific heat, neutron scattering, various magnetic resonance experiments, Mossbauer measurements, Kondo effect, and the like. Measurements have been extended also to alloys of many other systems besides those involving the platinum metals.

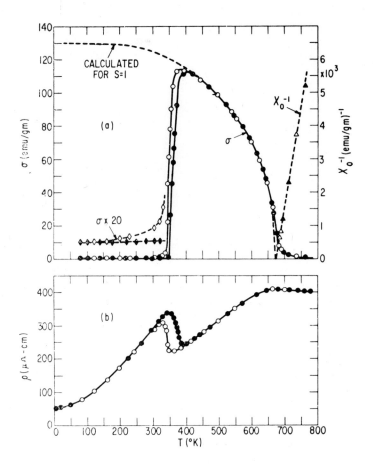

Journal of Applied Physics

Figure 6. (a) Magnetization (at 5 KOe) and inverse initial susceptibility, (b) Electrical resistivity of ordered FeRh for increasing (●) and decreasing (○) temperature (25)

The scope of this review precludes a description of these data, but the results of some of these experiments have shown weaknesses in the original band model approach (9, 20, 34). However, newer theoretical treatments (21, 22, 26) are providing more insight into the problem. The subject of localized magnetic states in dilute magnetic alloys is still in a state of very active research and theoretical development, as shown by the programs of even the most recent magnetic conferences, such as the Fifteenth Annual Conference on Magnetism and Magnetic Materials, Philadelphia, Pa., November 1969.

Ordered Alloys

In addition to the dilute alloys already discussed, there are a number of alloys of the metals of the platinum group with manganese, iron, cobalt, and nickel which have magnetic properties based on the formation of ordered structures at some particular composition.

Considering iron alloys first, the iron–rhodium system is an interesting example of magnetic ordering. Early magnetic measurements in the system (16) established that alloys of about 50 atomic % rhodium increased in magnetization as their temperature was raised through a critical value. Since 1960, a number of workers (23, 25, 36) have shown that this change is owing to a first-order antiferromagnetic (AFM) to ferromagnetic (FM) transition with increasing temperature, the transition temperature for a 52 atomic % rhodium being about 350°K, as shown in Figure 6 (25). X-ray diffraction studies have shown that the crystal structure above and below the transition temperature is an ordered CsCl-type, the transition being a uniform rapid volume expansion of about 1% with increasing temperature. Changes in lattice dimensions in this type of transformation result from the differences between the magnetoelastic expansion of the ferromagnetically ordered lattice and the contraction owing to the antiferromagnetic lattice. Measurements of the low-temperature specific heat of iron–rhodium alloys and iron–rhodium–palladium alloys and of entropy changes of the transition have shown that the difference in the energies of the FM and AFM state is relatively small. This suggests a model in which below the transition temperature in the AFM state, rhodium atoms, by symmetry, have no net exchange field exerted by the iron atoms and the energy is dominated by iron–iron interactions. Above the transition temperature, in the FM state, the rhodium atoms are polarized by an exchange field which induces a significant local moment on the rhodium atoms. This introduces another energy term which, if the susceptibility of the rhodium atoms is large enough, will account for the change to the FM state (23).

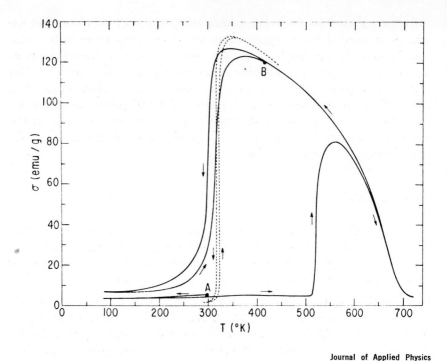

Figure 7. Magnetization in a 12.5 KOe field of FeRh filings. The dashed line is for the bulk alloys (27).

An interesting effect in iron–rhodium alloys is noted in cold-working the alloy. Filing an alloy to make powder is a convenient way of cold-working. Magnetic measurements on filings are shown in Figure 7 (27). Starting at point A, the alloy is cooled to 70°K and then heated to 500°K. It is evident that in the cold-worked alloy there is no trace of the AFM to FM transition in this range. Between 500° to 700°C, a normal Curie point behavior emerges and, on cooling, the first-order transformation reappears. X-ray diffraction measurements on the filings indicated that in the as-filed condition the filings have a disordered fcc structure, but quenching from temperatures as high as 1400°C did not result in the appearance of the fcc structure.

First-order transitions also exist for the compositions Pd_3Fe and Pt_3Fe, based on the Cu_3Au-type structure. In the ordered state, the former alloy is ferromagnetic and the latter is antiferromagnetic. Intermediate compositions based on $Fe(Pd,Pt)_3$ have shown the existence of a state at low temperatures in which, still based on the ordered Cu_3Au structure, the iron moments move into a "canted" structure as shown in Figure 8 (24). For the composition $FePd_{1.6}Pt_{1.4}$, the transition from a simple ferromagnetic state to this ferrimagnetic state with canted iron moments occurs at about 140°K.

The alloy Pt_3Fe with added iron is an interesting case by itself (1). The ordered alloy is antiferromagnetic with ferromagnetic sheets of iron atoms arranged on (110) planes antiferromagnetically along the [001] axis. As the iron content is increased over the stoichiometric 25 atomic %, a different AFM structure appears with the sheets arranged on (100) planes. At 30 atomic % iron, this structure predominates. Between these two extremes, both types of structure coexist. From neutron diffraction experiments, it is concluded that phase coherence of the two structures occurs over many unit cells and that there is an intertwining of the two structures rather than separation into domains of one or the other structure.

The platinum–cobalt system is especially important in that it has the only precious metal alloy ever to be developed for practical use of its magnetic properties. This alloy is based on stoichiometric or near-stoichiometric PtCo. It is distinguished by a very high-energy product, a relatively low remanence, and high coercivity. Fields of 30,000 oersteds or higher are required for saturation and PtCo magnets are regularly produced with energy products greater than 9 million gauss-oersteds, coercive forces of 4300 oersteds, and remanence near 6400 gauss. FePt

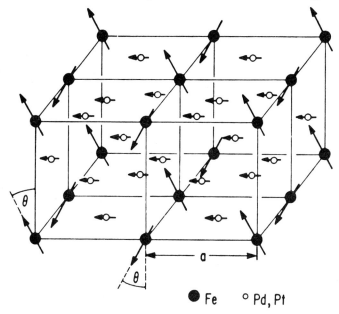

Journal of Applied Physics

Figure 8. Low-temperature canted-ferrimagnetic structure of $FePd_{1.6}Pt_{1.4}$ (24)

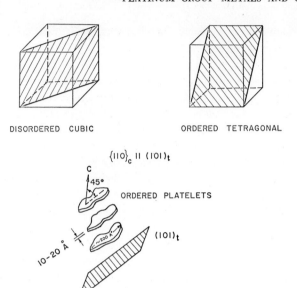

Figure 9. Crystallographic relationships between disordered and ordered PtCo

is isomorphous to PtCo and presumably shares magnetic hardening mechanisms. It has never been used practically because its coercivity and energy products are lower than those of PtCo.

PtCo is an ordering alloy, forming a face-centered-tetragonal CuAu-type structure ($c/a \sim 0.98$) from a disordered face-centered cubic lattice (*31*). Maximum coercive force and maximum energy product are achieved at less than complete order, and at different stages in the approach to complete order. Ordering starts in the disordered alloy with the formation of a system of ordered platelets, each containing a tetragonal c-axis parallel to one of the original orthogonal cube axes. These are (110) platelets; that is, they are not parallel to the "cube faces" in the disordered material. In any given region, primarily as an accommodation to strain, only two of the three possible platelet orientations occur (*8*). Figure 9 summarizes the crystallographic aspects of this system. Electron microscopic and field ion microscopic work has shown that maximum coercivity is achieved in the tetragonal phase with platelet width of 200–500 angstroms and with platelet thickness of about 20 angstroms (*30, 33*). The direction of easy magnetization (c-axis) is not in the plane of the platelet. The degree of transformation is far from complete at maximum properties, perhaps as little as 50% transformation by volume (*32*).

The saturation magnetization of the disordered alloy is 43.5 gauss cm^3/gm at 30,000 oersteds (*14*). The maximum internal magnetization

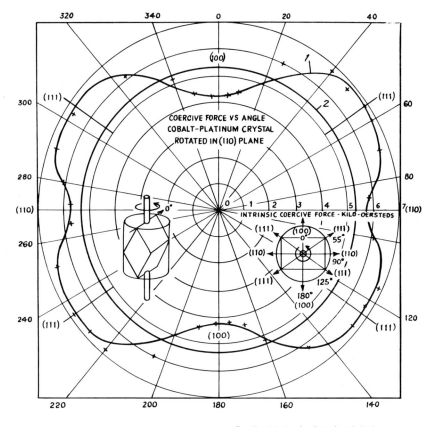

Figure 10. Magnetic anisotropy in a PtCo single crystal (37)

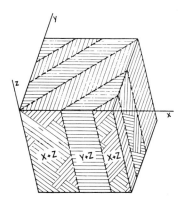

Figure 11. Crystallite distribution in ordered PtCo (8)

of the ordered phase has been estimated as 35–40% lower than that of the disordered phase (*14*). The average magnetic moment of disordered PtCo at room temperature is 1.04 bohr magnetons with a spherical distribution of moment (*8*). The easy magnetization direction in the ordered phase corresponds to its c-axis and the anisotropy constant is approximately 50 million erg/cm^3 (*8*).

Single-crystal work by Walmer (*37*) (on a fully-ordered crystal) showed a maximum of the coercive force in the 111 direction and minima in the 110 and 100 directions, the 100 minimum being the lower of the two (Figure 10). Work by Brissonneau and coworkers (*8*) on the distribution of platelets in completely-ordered PtCo has led to a model shown in Figure 11, wherein each zone shown in the figure contains a fully-developed network of (110) platelets oriented in two of the three possible orthogonal directions.

The zone model proposed by Brissonneau is consistent not only with his own field ion microscopic observations, but also with larger scale and often puzzling phenomena observed. PtCo magnets, for instance, form Bitter patterns on a scale quite easily visible under a light microscope. Bitter patterns do not fit the picture of a "fine-particle" hardening mechanism unless they also describe a larger scale phenomenon, such as Brissonneau zone boundaries.

Because of the sensitivity of the platelets to strain energy, their nucleation should be sensitive to the working history of the billets from which the alloy specimen is made, and, not surprisingly, PtCo responds well to cold working before heat treating. Work by Shimizu and Hashimoto (*35*), who tempered PtCo under elastic stress far below its elastic limit, has shown that the alloy develops considerably lower coercivity

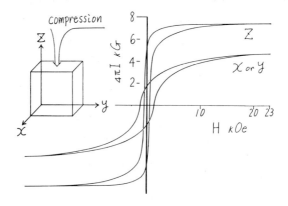

Journal of Applied Physics

Figure 12. Effects of compressive and tensile stresses on PtCo hysteresis curves (*35*)

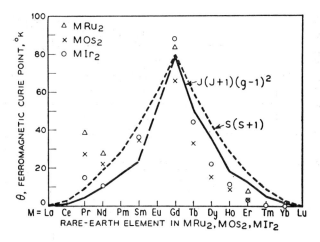

Figure 13. Curie temperatures for MX_2 compounds (M = rare earth, X = Ru, Os, or Ir) (7)

(and higher remanence) in the direction of the applied compressive stress than at right angles to it (Figure 12). They also report the converse to be true when tensile stress is applied.

Platinum Metal–Rare Earth Alloys

The magnetic moments of rare earth elements are caused by the unpaired electrons in their 4f shells. These shells are shielded by the outer shells so that chemical bonding has relatively little effect on the magnetic moments of these elements. The rare earths form a series of Laves phases of composition MB_2 (M = rare earth, B = precious metal) which share the characteristic of ferromagnetic coupling at low temperatures (7). Figure 13 shows the Curie temperatures of a series of com-

Table II. Ferromagnetic Curie Points[a]

Compound	θ, °K	Compound	θ, °K
$PrRu_2$	40	$NdRu_2$	35
$PrRh_2$	8.6	$NdRh_2$	8.1
$PrOs_2$	>35	$NdIr_2$	11.8
$PrIr_2$	18.5	$NdPt_2$	6.7
$PrPt_2$	7.9		
$GdRh_2$	>77		
$GdIr_2$	>77		
$GdPt_2$	>77		

[a] Reprinted from *Acta Crystallographica* (12).

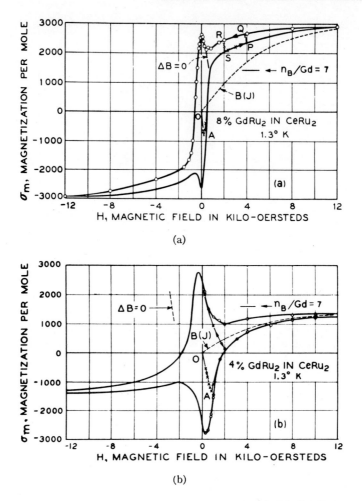

Journal of Applied Physics

Figure 14. Hysteresis loops for 2 $GdRu_2$–$CeRu_2$ alloys showing (a) superconductivity and ferromagnetism, (b) superconductivity alone (4)

pounds where B = Ru, Os, and Ir. These compounds form in the cubic $MgCu_2$ (C15) or the hexagonal $MgZn_2$ (C14) structures (7). The Curie point is highest with compounds containing Gd and decreases with larger or smaller atomic numbers of the rare earths (Figure 13). Similar results have been obtained with B = Rh or Pt (12) (Table II). The variation in Curie temperatures results principally from the interaction between the spins of the 4f shells of the rare earths and the conduction electrons.

Pseudobinary alloys of ($CeRu_2$–$GdRu_2$) (4), (YOs_2–$GdOs_2$) (29), ($GdRu_2$–$ThRu_2$) (6), ($GdOs_2$–$LaOs_2$) (5), and perhaps some others show simultaneous ferromagnetic and superconducting properties over a limited range of composition. The system $GdRu_2$ in $CeRu_2$ has been studied extensively. This system shows a superconducting transition when the concentration of $GdRu_2$ is less than 10 atomic %. Alloys containing more than 6% $GdRu_2$ are ferromagnetic with Curie temperatures increasing with concentration (4). Figure 14 shows hysteresis loops for 8 and 4% $GdRu_2$ alloys, the latter being superconducting and the former being simultaneously superconducting and ferromagnetic. Both loops show the exclusion of field characteristic of superconducting materials,

Table III. Curie Temperature and Coercive Force of R_5Pd_2 Compounds (3)

Compound	θ, °K	Hc (at $4.2°K$) oe
Gd_5Pd_2	335	–
$Tb_{5.10}Pd_{1.90}$	~30	12,800
$Dy_{5.07}Pd_{1.93}$	~25	9,700
$Ho_{5.04}Pd_{1.96}$	~10	1,300

i.e., negative slope on initial magnetization and negative slope near remanence in the first quadrant. The normal or nonferromagnetic superconductor exhibits a "remanence" attributed to "frozen-in" flux. The magnetization curve for the ferromagnetic (8%) alloy lies well above that calculated for a paramagnetic material of the same Gd content, and the remanence is also well above that expected for a paramagnetic superconductor. It is questionable whether superconductivity and ferromagnetism exist in the same domains in a given specimen. The minor loop PQRS shows that some superconductivity still exists in parts of the alloy after superconductivity as a whole has been destroyed.

Compounds of nominal composition M_5B_2, where M is again a rare earth (Gd, Tb, Ho, Dy) and B a precious metal (Pt, Pd), show magnetizations close to those calculated from the moments of the rare earth atoms (3). Gd_5Pd_2 is a soft magnetic material with a coercive force less than 100 oersteds but with a high saturation associated with the high moment of Gd (3). Table III summarizes the permanent magnetic properties of the palladium alloys. The platinum alloys are isostructural and have the same Curie temperatures (19). Energy products for Tb_5Pd and Dy_5Pd at 4.2°K are 20×10^6 and 26×10^6 gauss-oersteds, respectively (3). While these energy products are higher than platinum–cobalt, the low Curie temperatures preclude use of these compounds in the usual platinum–cobalt applications.

Literature Cited

(1) Bacon, G. E., Crangle, J., *Proc. Roy. Soc.* **1963**, A272, 387.
(2) Barton, E. E., Claus, H., private communication.
(3) Berkowitz, A. E., Holtzberg, F., Methfessel, S., *J. Appl. Phys.* **1964**, 35, 1030.
(4) Bozorth, R. M., Davis, D. D., *J. Appl. Phys.* **1960**, 31, 321S.
(5) Bozorth, R. M., Davis, D. D., Williams, A. J., *Phys. Rev.* **1960**, 119, 1570.
(6) Bozorth, R. M., Matthias, B. T., Davis, D. D., *Proc. Intern. Conf. Low Temp. Phys.*, 7th, Toronto, Canada, 1960, **1961**, p. 385.
(7) Bozorth, R. M., Matthias, B. T., Suhl, H., Corenzwit, E., Davis, D. D., *Phys. Rev.* **1959**, 115, 1595.
(8) Brissonneau, P., Blanchard, A., Schlenker, M., Laugier, J., *J. Appl. Phys.* **1969**, 39, 1266.
(9) Brog, K. C., Jones, W. H., Booth, J. G., *J. Appl. Phys.* **1967**, 38, 1151.
(10) Cape, J. A., Hake, R. R., *Phys. Rev.* **1965**, 139, A142.
(11) Clogston, A. M., et al., *Phys. Rev.* **1962**, 125, 541.
(12) Compton, V. B., Matthias, B. T., *Acta Cryst.* **1959**, 12, 651.
(13) Doclo, R., Foner, S., Narath, A., *J. Appl. Phys.* **1969**, 40, 1206.
(14) Dunaev, F. N., Kalinin, V. M., Kryokov, I. P., Maisinovich, V. I., *Fiz. Metal i Metaloved.* **1965**, 20, 460.
(15) *Engelhard Ind. Tech. Bull.* VI, No. 3, December, 1965.
(16) Fallot, M., Hocart, R., *Rev. Sci.* **1939**, 77, 498.
(17) Foner, S., Doclo, R., McNiff, E. J., Jr., *J. Appl. Phys.* **1968**, 39, 551.
(18) Geballe, T. H., et al., *J. Appl. Phys.* **1965**, 37, 1181.
(19) Holtzberg, F., Methfessel, S. J., U. S. Patent **3,326,637** (1967).
(20) Knapp, G. S., *J. Appl. Phys.* **1967**, 38, 1267.
(21) Knapp, G. S., *Phys. Letters* **1967**, 25A, 114.
(22) Kondo, J., *Phys. Rev.* **1968**, 169, 437.
(23) Kouvel, J. S., *J. Appl. Phys.* **1966**, 37, 1257.
(24) Kouvel, J. S., Forsyth, J. B., *J. Appl. Phys.* **1969**, 40, 1359.
(25) Kouvel, J. S., Hartelius, C. C., *J. Appl. Phys.* **1962**, 33, 1343.
(26) Lederer, P., Mills, D. L., *Phys. Rev.* **1968**, 165, 837.
(27) Lommel, J. M., Kouvel, J. S., *J. Appl. Phys.* **1967**, 38, 1263.
(28) Low, G. G., Holden, T. M., *Proc. Phys. Soc.* **1966**, 89, 119.
(29) Matthias, B. T., U. S. Patent **2,970,961** (1961).
(30) Mishin, D. D., Greshishkin, R. M., *Phys. Status Solidi* **1967**, 19, K1-K3.
(31) Newkirk, J. B., Geisler, A. H., Martin, D. L., Smoluchowski, R., *J. Metals* **1950**, 188, 1249.
(32) Rabinkin, A. G., Tyapkin, Yu. D., Yamaleev, K. M., *Fiz. Metal i Metaloved.* **1965**, 19, 360.
(33) Ralph, B., Brandon, D. G., *Proc. European Conf. Electron Microscopy*, 3rd, Prague **1964** (A) 303.
(34) Sarachik, M. P., *Phys. Rev.* **1968**, 170, 679.
(35) Shimizu, S., Hashimoto, E., *J. Appl. Phys.* **1968**, 39, 2369.
(36) Tu, P., et al., *J. Appl. Phys.* **1969**, 40, 1368.
(37) Walmer, M. L., *Engelhard Ind. Tech. Bull.* **1962**, 2, 117.
(38) Walstedt, R. E., Sherwood, R. C., Wernicke, J. H., *J. Appl. Phys.* **1968**, 39, 555.
(39) Weiss, R. J., "Solid State Physics for Metallurgists," p. 323, Pergamon, New York, 1963.

RECEIVED January 16, 1970.

2

Platinum Metal Chalcogenides

AARON WOLD

Brown University, Providence, R. I. 02912

> A number of binary platinum metal chalcogenides show interesting electrical and magnetic properties. These are classified according to structure types and their properties related to the formal valence of the platinum metal. A simple one-electron model proposed by Goodenough to explain the properties of the first row transition metal chalcogenides can be applied to the corresponding platinum metal compounds. This theory is successful in correlating new experimental data on these compounds—e.g., crystal structure, resistivity, Seebeck coefficient, and magnetic susceptibility.

The electrical and magnetic properties of simple binary platinum metal oxides (15, 19), as well as a number of ternary compounds (4), have shown that these materials can be of considerable interest to both solid state chemists and physicists. A number of the platinum metal oxides are remarkably good electrical conductors and, in addition, magnetic ordering has been observed (4) for several perovskites containing ruthenium. Many of these compounds can be prepared in the form of single crystals (18), making them suitable for both transport and magnetic measurements.

The platinum metal chalcogenides in general are easier to prepare than the corresponding oxides. Whereas special conditions of temperature and pressure are required to prepare many of the oxides, the platinum metals react most readily with S, Se, and Te. A number of additional differences concerning the chemistry of the chalcogenides and the oxides are summarized as follows: The metal–sulfur (selenium, tellurium) bond has considerably more covalent character than the metal–oxygen bond and, therefore, there are important differences in the structure types of the compounds formed. Whereas there may be considerable similarity between oxides and fluorides, the structural chemistry of the sulfides tends to be more closely related to that of the chlorides. The latter compounds form as primarily layer-type structures, and a number of important structure types are found in sulfides that have no counterpart among the oxide structures—e.g., pyrite, marcasite, nickel arsenide. Finally, many of the platinum metal chalcogenides behave like alloys; the elements do not appear to have their normal valencies and, in addition, the compounds have a wide range of composition and show metallic luster, reflectivity, and conductivity.

Goodenough (10) has accounted for the metallic properties of a number of first row transition metal chalcogenides, and these ideas are also applicable to the platinum metal chalcogenides. He has shown that the formation of conduction bands is possible as a result of strong covalent mixing between the cation e_g and anion s,p_σ orbitals. This mixing allows sufficient interaction between octahedrally coordinated cations on opposite sides of an anion so that the conditions for localized d-electrons break down. The resulting d-like collective-electron bands, which are antibonding with respect to the anion array, are designated σ^* bands. Metallic behavior can occur when these bands exist and are partially occupied by electrons. Metallic conduction may occur also if conduction bands are formed by the direct overlap of transition metal $t_{2g} - t_{2g}$ orbitals. Here again, these bands must be partially occupied. Rogers et al. (19) have applied the Goodenough model to account for the electrical behavior of a number of platinum metal oxides. These concepts may also be applied to the platinum metal chalcogenides.

It will not be possible, in this paper, to deal with all of the platinum metal chalcogenides. Instead, a number of examples will be chosen and their electrical as well as magnetic properties correlated with the atomic positions in the various structures formed. The first group of compounds to be discussed crystallize with the pyrite structure, which is shown in Figure 1. This structure is similar to the NaCl structure if we replace Na by Fe and each Cl by an S_2 group. However, the S–S distance within

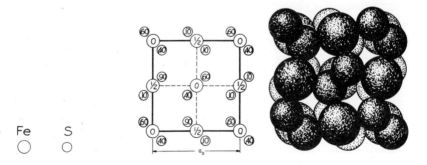

R. W. G. Wyckoff, "Crystal Structures," Wiley

Figure 1. Pyrite structure (20)

Table I. Platinum Metal Compounds with the Pyrite Structure

Compounds — Properties

Low Spin d^6 S = 0

Compounds				Properties
RuS_2,	$RuSe_2$,	$RuTe_2$		Semiconductor
OsS_2,	$OsSe_2$,	$OsTe_2$		Semiconductor
RhPS,	RhAsS,	$RhSbS$,	RhBiS	Semiconductor
RhPSe,	RhAsSe,	RhSbSe,	RhBiSe	Semiconductor
—	RhAsTe,	RhSbTe,	RhBiTe	Semiconductor
IrPS,	IrAsS,	$IrSbS$,	IrBiS	Semiconductor
IrPSe,	IrAsSe,	IrSbSe,	IrBiSe	Semiconductor
—	IrAsTe,	IrSbTe,	IrBiTe	Semiconductor
$Ni_{0.2}Pd_{0.5}As_2$,		$PdAs_2$,	$PdSb_2$	Metallic
PtP_2,	$PtAs_2$,	$PtSb_2$		Semiconductor
$PtBi_2$				Metallic
$RhS_{\sim 3}$,	$RhSe_{\sim 3}$,	$RhTe_{\sim 3}$		Semiconductor
$IrS_{\sim 3}$,	$IrSe_{\sim 3}$			Semiconductor
$IrTe_{\sim 3}$				Metallic

$$d^7 \ S = \frac{1}{2} \pm 0$$

Compounds					Properties
$RhSe_2$,	$RhTe_2$				
IrS_2					
PdAsS,	PdSbS,	PdAsSe,	PdSbSe,	PdBiSe	Metallic
PdSbTe,	PdBiTe,	PtAsS,	PtSbS,	PtSbSe	Metallic
PtBiSe,	PtSbTe,	PtBiTe			Metallic

the S_2 group is such that each iron is surrounded by a deformed anion octahedron, and the anions have three cation and one anion neighbors, which together form a distorted tetrahedron. A number of compounds that crystallize with the pyrite structure are listed in Table I. For many of them, the transition metal has the low-spin d^6 state. These compounds are semiconductors because the conduction band (σ^*, formed via e_g orbitals) is empty. An apparent exception is $IrTe_{\sim 3}$, which is reported (17) to be a superconductor. As one proceeds from sulfur to selenium to tellurium (as anions in an isostructural series of compounds) there is a narrowing of the band gap, and for some series of compounds, this is accompanied by a marked change in the transport properties.

The metallic properties observed for the d^7 compounds listed in Table I are also consistent with the Goodenough model. The rhodium–selenium system is of particular interest and demonstrates clearly the important relationships between structure and transport properties. Cations may be removed from the superconducting compound ($T_c = 6°K$), $RhSe_2$. The pyrite structure is maintained as the e_g band is gradually emptied, and at the composition $Rh_{2/3}Se_2$ ($RhSe_3$), all cations are trivalent—i.e., have the configuration $4d^6$. It is not known yet if the ideal composition, $RhSe_3$, can be obtained. A homogeneity range is reported (9) from $RhSe_{1.8}$ to $RhSe_{2.7}$, and we know only that $RhSe_3$ must be the upper limit. However, RhS_3 and $IrSe_3$ are diamagnetic (12) and semiconducting, and stoichiometric $RhSe_3$ should show semiconducting behavior also.

Another interesting compound is $IrSe_2$, which should show metallic behavior since the Ir appears to have a d^7 configuration. However, the $IrSe_2$-type compounds are diamagnetic semiconductors, and their structures resemble that of marcasite. The marcasite structure in Figure 2 shows hexagonal close-packed anion layers with half the layers of the octahedral holes filled. The $IrSe_2$ structure is shown in Figure 3, and the relationship to the marcasite structure is seen readily by comparing the octahedral chains. The $IrSe_2$ structure evolves from an anion stacking ABABACAC. Units of the marcasite structure are stacked together in such a way that only half the anions form pairs, and the cations are indeed trivalent $Ir(III)Se \cdot (Se_2)_{1/2}^{2-}$. The cation, therefore, actually

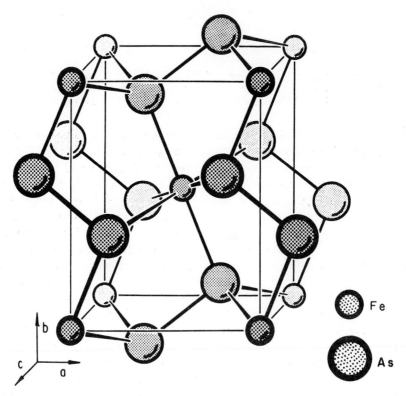

F. Hulliger, "Structure and Bonding," Springer-Verlag

Figure 2. Marcasite structure (13, p. 94)

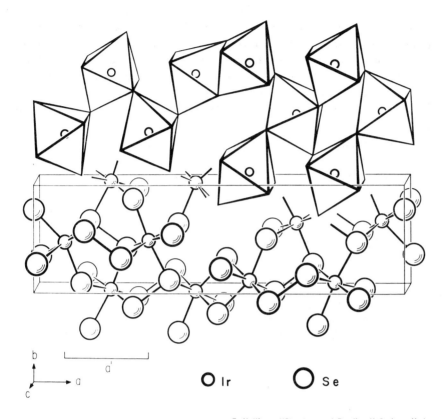

F. Hulliger, "Structure and Bonding," Springer-Verlag

Figure 3. IrSe$_2$ structure (13, p. 100)

has a d^6 configuration consistent with the semiconducting, diamagnetic properties observed.

The structure of PdSe$_2$, an elongated pyrite, is shown in Figure 4. The square-planar coordination is achieved by elongating the anion octahedron. This decrease in symmetry results in a strong splitting of the e_g levels (d_{z^2} stabilized relative to $d_{x^2-y^2}$), and the 8 d-electrons of palladium just fill all available d-levels through the lower d_{z^2} level. PdS$_2$, PdSSe, and PdSe$_2$ are the only known compounds of this structure type and are (11) diamagnetic semiconductors. When the distortion of the anion octahedra in PdS$_2$ is reduced by pressure, the semiconductor transforms into a metal before a change to the pyrite structure occurs (1).

A number of platinum metal chalcogenides (PtS$_2$, PtSe$_2$, PtSeTe, PtTe$_2$) crystallize in the CdI$_2$ structure (Figure 5). The cations occupy half the holes of a hexagonal close-packed array, such that filled and empty layers alternate to give a sandwich-like structure. The cadmium atoms are symmetrically surrounded by six halogen atoms at the corners of an octahedron, whereas the three cadmium neighbors of each halogen atom all lie on one side of it. Since there are no bonds between like atoms, a tetravalent cation is needed to make a CdI_2-type chalcogenide nonmetallic—e.g., PtS_2. Narrowing of the band gap is observed for this series of platinum metal chalcogenides as one proceeds from PtS_2 to $PtTe_2$. Whereas PtS_2 and $PtSe_2$ are semiconductors, PtSeTe and $PtTe_2$ are metallic.

Several platinum metal chalcogenides of the type MX crystallize with the NiAs structure. They include RhSe, RhTe, and PdTe. This structure is shown in Figure 6. Each cation is coordinated by six anion neighbors

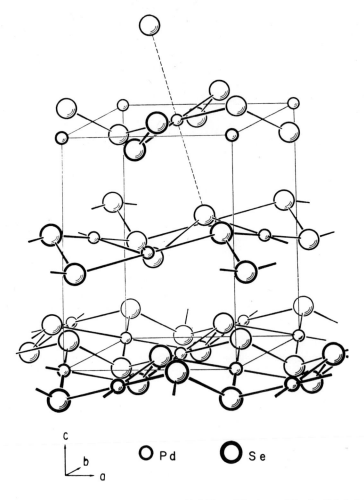

F. Hulliger, "Structure and Bonding," Springer-Verlag

Figure 4. $PdSe_2$ structure (13, p. 108)

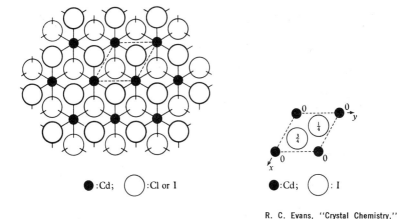

●:Cd; ○:Cl or I   ●:Cd; ○:I

R. C. Evans, "Crystal Chemistry,"
Cambridge University Press

Figure 5. CdI$_2$ structure (7)

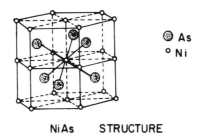

NiAs STRUCTURE

Figure 6. Nickel arsenide structure

at the corners of a distorted octahedron, but the six cation neighbors of an anion are located at the corners of a trigonal prism. Each cation also has two other cation neighbors only slightly more remote than its anion neighbors. The structure may be considered a filled CdI$_2$ structure, with all of the octahedral sites in the close-packed hexagonal anion sublattice now occupied by cations. An interesting group of ternary compounds may be formed in which one-fourth of the sites are empty. This latter structure type was studied first by Jellinek (*14*) and usually is designated as the Cr$_3$S$_4$ structure.

Jellinek has pointed out also that the tetragonal structure of cooperite, PtS, is related to NiAs. In diamagnetic PtS(d^8), four d-orbitals are doubly occupied and one is empty. Therefore, two anions from a hypothetical octahedron have been removed, and the cation is in a square-planar coordination; the anion is coordinated by four cations that form a deformed tetrahedron. The cooperite structure is shown in Figure 7. The relationship between the NiS and PtS structure is similar to that between

Table II. Room-Temperature Structural and Electrical

Compound	System	$a\ (\pm\ 0.005)$, Å	$b\ (\pm\ 0.005)$, Å
$CrRh_2Se_4$	Monoclinic	6.278	3.612
$CoRh_2Se_4$	Monoclinic	6.269	3.644
$NiRh_2Se_4$	Monoclinic	6.280	3.648
$CrRh_2Te_4$	Monoclinic	6.841	3.951
Rh_3Te_4	Monoclinic	6.812	3.954
$CoRh_2Te_4$	Trigonal	3.953	—
$NiRh_2Te_4$	Trigonal	3.966	—

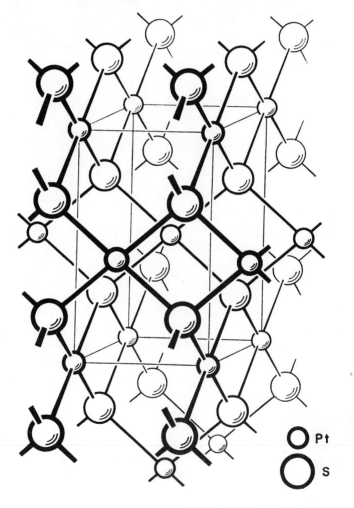

F. Hulliger, "Structure and Bonding," Springer-Verlag

Figure 7. Cooperite (PtS) structure (13, p. 166)

Data for Ternary Rhodium Chalcogenides

c ($\pm$ 000.5), Å	β ($\pm$ 0.05), Deg	Electrical Resistivity, Ohm-Cm
11.25	92.47	1×10^{-3}
10.81	92.15	—
10.82	92.22	5×10^{-4}
11.40	91.61	1×10^{-3}
11.23	92.55	1×10^{-4}
5.429	—	—
5.457	—	3×10^{-4}

NiS_2 and PdS_2. There is an elongation of the anion octahedra and a change from a high-spin to a diamagnetic low-spin state for the d^8 cations.

In addition to the simple binary platinum metal chalcogenides, of which we have chosen only a few examples, a number of interesting ternary compounds have been prepared. Blasse (2, 3) has reported the preparation and properties of several rhodium thiospinels. The copper compound, $CuRh_2S_4$, shows a weak, nearly temperature-independent paramagnetism of 320×10^{-6} emu/g-mole. The structure is an undistorted cubic spinel, indicating that the Cu^{2+} on the tetrahedral site does not show its usual tendency to distort to lower symmetry via a Jahn-Teller mechanism. According to the Goodenough model, these properties are expected for $CuRh_2S_4$ if the unpaired $3d$ electrons from the Cu^{2+} are delocalized and occupy a σ^* conduction band formed as a result of covalent mixing between the $Cu(t_{2g})$ and sulfur (p_π) orbitals. For $CoRh_2S_4$, Blasse reported that magnetic susceptibility showed Curie-Weiss behavior between 450°–800°K. The effective moment was $4.3\mu_B$ (the spin only moment for Co^{2+} is $3.9\mu_B$). At 400°K, antiferromagnetic ordering was observed. The other thiospinels were apparently too difficult to prepare as single-phase materials. Plovnick and Wold (16) reported the

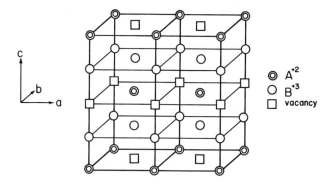

Figure 8. Cr_3S_4 structure

preparation of the compounds MRh_2X_4 (M = Cr, Co, Ni; X = Se, Te) and Rh_3Te_4, which have the Cr_3S_4-type structure (space group $I2/m$), except for $CoRh_2Te_4$ and $NiRh_2Te_4$, which are trigonal (space group $P\bar{3}m1$). The monoclinic structure shown in Figure 8 is intermediate between the NiAs and CdI_2 types. The transition metal cations, A and B, occupy three-fourths of the octahedral sites between the layers of hexagonal close-packed anions. The packing sequence is such that B^{3+} layers alternate with layers containing the A^{2+} cations and vacancies. Ideally, the A^{2+} cations are ordered with respect to the vacancies. When the vacancies are randomly arranged in A^{2+} layers, trigonal symmetry results. The crystal class to which a given AB_2X_4 compound belongs is indicative of the vacancy arrangement in that compound.

The properties of a number of ternary rhodium chalcogenides are given in Table II. $CoRh_2Se_4$ and $NiRh_2Se_4$ are monoclinic, while $CoRh_2Te_4$ and $NiRh_2Te_4$ are trigonal. Co_3Se_4 and Ni_3Te_4 are also trigonal. The monoclinic structures observed for $CrRh_2Se_4$ and $CrRh_2Te_4$ are consistent with reports (5, 6) that both Cr_3Se_4 and Cr_3Te_4 are monoclinic. In chalcogenides of this type, both the size and d-electron configuration of the cations, as well as the size and polarizability of the anion, are important in determining the relative stability of the ordered *vs.* random vacancy structure for a given composition.

Rh_3Te_4 also has the monoclinic Cr_3S_4-type structure (*15*). Geller (*8*) determined the crystal structure of RhTe and $RhTe_2$ but did not investigate intermediate compositions. Plovnick and Wold (*16*) have shown that in the rhodium–tellurium system at the composition Rh_3Te_4, ordering of the vacancies occurs with a consequent lowering of the symmetry to monoclinic.

The Goodenough model (*10*) can account for the electrical properties of ternary rhodium chalcogenides of the type ARh_2X_4. The octahedrally coordinated $Rh^{3+}(4d^6)$ has the low-spin configuration, and no contribution to the conductivity is made either by direct interaction of the cation t_{2g} orbitals or by indirect e_g-anion s,p_σ interaction. This is not the case, however, for the M cations in the MRh_2X_4 compounds. For example, $Ni^{2+}(t_{2g}^6 e_g^2)$ may contribute to metallic conductivity *via* formation of partially-filled σ^* bands as a result of nickel e_g-anion s,p_σ interactions. Similar considerations apply to the other ternary rhodium chalcogenides.

Literature Cited

(1) Bither, T. A., private communication.
(2) Blasse, G., *Phys. Letters* **1965**, 19 No. 2, 110.
(3) Blasse, G., Schipper, D. J., *Phys. Letters* **1963**, 5 No. 5, 300.

(4) Callaghan, A., Moeller, C. W., Ward, R., *Inorg. Chem.* **1966,** 5 No. 9, 1572.
(5) Chevreton, M., Berodias, G., *Compt. Rend.* **1965,** 261, 1251.
(6) Chevreton, M., Bertaut, F., *Compt. Rend.* **1961,** 253, 145.
(7) Evans, R. C., "Crystal Chemistry," 2nd ed., p. 151, Cambridge University Press, London, 1964.
(8) Geller, S., *J. Am. Chem. Soc.* **1955,** 77, 2641.
(9) Geller, S., Cetlin, B. B., *Acta Cryst.* **1955,** 8, 272.
(10) Goodenough, J. B., *Proc. Colloq. Intern. Orsay* **1965;** *C.N.R.S.* **1967,** 157, 263.
(11) Hulliger, F., *J. Phys. Chem. Solids* **1965,** 26, 639.
(12) Hulliger, F., *Nature* **1964,** 204, 644.
(13) Hulliger, F., "Structure and Bonding," Vol. 4, Springer-Verlag, New York, 1968.
(14) Jellinek, F. K., *Acta Cryst.* **1957,** 10, 620.
(15) Marcus, S. M., Butler, S. R., *Phys. Rev. Letters* **1968,** 26A, 518.
(16) Plovnick, R. H., Wold, A., *Inorg. Chem.* **1968,** 7, 2596.
(17) Raub, C. J., Compton, V. B., Geballe, T. H., Matthias, B. T., Maita, J. P., Hull, G. W., *J. Phys. Chem. Solids* **1965,** 26, 2051.
(18) Rogers, D. B., Butler, S. R., Shannon, R. D., *Inorg. Syn.* 13, to be published.
(19) Rogers, D. B., Shannon, R. D., Sleight, A. W., Gillson, J. L., *Inorg. Chem.* **1969,** 8, 841.
(20) Wyckoff, R. W. G., "Crystal Structures," Vol. 1, 2nd ed., Wiley, New York, 1963.

RECEIVED December 16, 1969. This work has been supported by ARPA and the U. S. Army Research Office, Durham, N. C.

3

Synthesis and Crystal Chemistry of Some New Complex Palladium Oxides

OLAF MULLER and RUSTUM ROY

Materials Research Laboratory, The Pennsylvania State University, University Park, Pa. 16802

> *The synthesis and crystal chemistry of six new complex palladium oxides are described. The black $PbPdO_2$ has been tentatively indexed on a hexagonal basis with $a_o = 10.902 A$ and $c_o = 4.654 A$. Sr_2PdO_3 is body-centered orthorhombic with $a_o = 3.970 A$, $b_o = 3.544 A$, and $c_o = 12.84 A$; the structure proposed for this compound is closely related to the K_2NiF_4 structure. Three new black compounds of the $MePd_3O_4$ type have been synthesized. All three have the cubic $Na_xPt_3O_4$ structure with unit cell constants $a_o = 5.826 A$ for $SrPd_3O_4$, $a_o = 5.747 A$ for $CaPd_3O_4$, and $a_o = 5.742 A$ for $CdPd_3O_4$. In the system Mg–Pd–O, a spinel phase is formed ($a_o = 8.501 A$) whose stoichiometry is believed to be Mg_2PdO_4.*

In our recent work on the systems Rh–O, Pt–O (5, 7), and Au–O (5, 6) at high oxygen pressures, we prepared new compounds in all three systems. We conducted a similar study in the system Pd–O, but were not able to synthesize any new binary oxides of palladium. The divalent oxide, PdO, appears to be stable even at high oxygen pressures.

Our work on the platinum oxides was recently extended to some ternary systems, and we have already reported on the inverse spinels Mg_2PtO_4, Zn_2PtO_4 (10), on Cd_2PtO_4 with the Sr_2PbO_4 structure (9), on the tetragonal $Cu_{1-x}Pt_xO$, and the orthorhombic $CuPt_3O_6$ (8). Compounds of the type $MePt_3O_6$ with Me = Cd, Zn, Mg, Ni were synthesized at 800°–900°C and oxygen pressures near 200 atm. These phases are described in greater detail by Hoekstra, Siegel, and Gallagher (2). We encountered some difficulty in preparing these compounds in a pure state. Apparently, the maximum temperatures attainable with our equipment ($\sim$900°C) are still too low for these reactions. We felt that it might be interesting to conduct a similar study of some MeO–palladium oxide systems. The present work is concerned with our study of the systems Me–Pd–O, where Me = Pb, Sr, Ca, Cd, and Mg.

Relatively little is known about anhydrous oxides of palladium. The only well-characterized simple oxide is PdO, which has the PtS structure with four coplanar Pd–O distances of 2.01A (4, 21). Recently, Guiot (1) presented x-ray evidence suggesting that a new palladium oxide surface compound is formed as an intermediate step when palladium metal is oxidized in air to PdO. However, the stoichiometry of this compound is unknown, since it can be obtained only as a minor constituent, with major amounts of Pd metal and PdO. Higher anhydrous oxides (e.g., PdO_2) have been reported, but their existence has never been firmly established.

Relatively few ternary palladium oxides have been reported in the literature. Various alkali metal–palladium oxides are known (14, 15, 16). Intermediate compounds have been found in several Ln_2O_3–PdO systems with Ln = La, Nd, Sm, Eu, Gd, and Dy (3). At high pressures, pyrochlores of the type $Ln_2Pd_2O_7$ can be prepared (19). Recently, the delafossite-like compounds $PdCoO_2$, $PdCrO_2$, and $PdRhO_2$ have been studied (18).

Experimental Techniques

Cold-seal stellite bombs and high oxygen pressures were used as described previously (6, 7, 8, 9, 10). An ordinary pot furnace was used for heat treatments carried out in air. As starting materials, purified palladium black (Fisher Scientific Co.) was mixed intimately with the other oxides or hydroxides [PbO, $Sr(OH)_2 \cdot 8H_2O$, CaO, CdO, $Mg(OH)_2$]. The mixtures generally were inserted into gold foils and heat treated under various conditions of temperature and oxygen pressure. The runs were quenched and the products examined by x-ray diffraction.

The x-ray diffraction patterns were made with a Picker diffractometer, using Ni-filtered CuK radiation and glass-slide mounts. To derive accurate unit cell constants, slow-scan x-ray patterns were internally standardized with NaCl, KCl, or Si. The integrated intensities were obtained by measuring the area under each peak with a planimeter.

The cation analyses were performed by conventional wet-chemical techniques. The oxygen content was determined by a reduction method and by neutron activation analysis.

Results

$PbPdO_2$. The black $PbPdO_2$ is prepared conveniently by heating a 1:1 PbO–Pd black mixture in air at about 600°–700°C for several days.

Table I. X-Ray Diffraction Powder Data for $PbPdO_2$

Hexagonal: $a_o = 10.902$A, $c_o = 4.654$A

d obs., A	d calc., A	I obs.	hkl
4.72	4.72	2.0	200
3.544	3.539	5.1	111
2.833	2.832	100.0	211
2.727	2.726	19.7	220
2.607	2.607	2.8	301
2.360	2.360	17.0	400
2.327	2.327	12.0	002
2.282	2.282	1.7	311
2.087	2.087	4.1	202
1.964	1.964	4.0	321
1.784	1.784	18.4	420
1.769	1.770	20.3	222
1.693	1.693	1.5	331
1.657	1.657	15.1	402
1.593	1.593	22.8	511
1.573	1.574	1.8	600
1.531	1.531	1.3	103
1.492	1.492	0.4	113
1.438	1.438	16.6	521
1.419 B	{1.423, 1.416}	29.7	{213, 422}
1.392	1.391	0.6	303
1.375	1.375	1.4	611
1.363	1.363	4.5	440
1.334	1.335	0.7	313
1.305 B	{1.309, 1.308, 1.304}	1.0	{620, 441, 602}
1.261	1.261	0.6	323
1.239	1.239	1.2	413
1.2075	1.2077	0.2	711
1.1984	1.1987	0.6	503
1.1779	{1.1802, 1.1798, 1.1759}	12.4	{800, 333, 442}
1.1634	1.1635	2.8	004
1.1523	1.1525	8.0	631
1.1444	1.1446	9.5	513
1.1296	1.1297	1.4	204
1.1195	1.1196	0.4	721
1.0972	1.0973	0.7	433
1.0828	{1.0830, 1.0827}	9.8	{640, 523}
1.0701	1.0701	3.7	224
1.0527	1.0526	10.7	802
1.0435	1.0436	3.1	404
1.0358	1.0356	6.8	731

An analysis for oxygen by neutron activation yielded 9.2% O (calculated: 9.26% O by weight). PbPdO$_2$ is apparently unstable at temperatures above ~820°C, where a mixture of PbO and PbPd$_3$ is obtained in place of PbPdO$_2$.

The x-ray powder pattern of PbPdO$_2$ (Table I) has been tentatively indexed on the basis of a hexagonal unit cell with $a_o = 10.902$A and $c_o = 4.654$A. This may represent only a pseudocell. No single crystal data are available, since the product could only be obtained as a fine powder. Table I indicates that several extinctions occur which are not characteristic for any hexagonal space group; e.g., for $hk0$, $h = 2n$ and $k = 2n$.

Sr$_2$PdO$_3$. A yellow-brown compound, Sr$_2$PdO$_3$, is formed when the appropriate starting mixture (in pellet form) is fired at 950°C in air for several days. The compound is thermally quite stable, even at 1200°C. It dissolves rapidly in dilute acids, and when brought in contact with water, it alters to an unidentified hydrated product. Analyses lead to the formula ~Sr$_{1.93}$Pd$_{1.00}$O$_{3+x}$ with x as high as 0.6. We have assumed an ideal formula of Sr$_2$PdO$_3$ for reasons discussed below.

Rotation and Weissenberg patterns were taken of a single crystal of Sr$_2$PdO$_3$. The single-crystal data lead to a body-centered orthorhombic unit cell. Powder data ($a_o = 3.970$A, $b_o = 3.544$A, $c_o = 12.84$A) are in good agreement with the single-crystal data. The only extinctions found were those for body-centering; thus, the four possible space groups are D_{2h}^{25}-*Immm*, D_2^8-*I222*, D_2^9-*I2$_1$2$_1$2$_1$*, and C_{2v}^{20}-*Imm2*.

Both symmetry and unit cell constants are similar to the corresponding values of K$_2$NiF$_4$ type compounds, as is shown in Table II. Indeed, we have found that a reasonable structure can be derived for Sr$_2$PdO$_3$ which is closely related to the K$_2$NiF$_4$ structure. In this proposed structure, the space group D_{2h}^{25}-*Immm* is assumed and the atoms are placed as follows:

4 Sr in (4i) : $00z$, $00\bar{z}$ + b.c. with $z = 0.355$

2 Pd in (2a) : 000 + b.c.

2 O in (2b) : $0\frac{1}{2}\frac{1}{2}$ + b.c.

4 O in (4i) : $00z$, $00\bar{z}$ + b.c. with $z = 0.16$

With this assumption, intensities for the powder pattern were calculated in the same manner as described previously (8). All temperature coefficients were assumed to be zero. The calculated intensities are listed in Table III and compare quite favorably with the observed intensities. The intensity discrepancy factor $R = 100 \ (\Sigma|I_o - I_c|)/(\Sigma \ I_o)$ was calculated as 6.6. In calculating the R-factor, the intensities which were too weak to be observed were assigned I obs. values of zero. The

Table II. Structural Data for Some Strontium–Noble Metal Oxides

Compound	Unit Cell Constants			Ref.	Symmetry Space Group Structure Type
	a_0, A	b_0, A	c_0, A		
Sr_2IrO_4	3.89		12.92	(12)	Tetragonal
Sr_2RhO_4	3.85		12.90	(13)	D_{4h}^{17}-$I4/mmm$
Sr_2RuO_4	3.870		12.74	(13)	K_2NiF_4 structure
Sr_2PdO_3	3.970	3.544	12.84	present work	Orthorhombic D_{2h}^{25}-$Immm$

Table III. X-Ray Diffraction Powder Data for Sr_2PdO_3

Orthorhombic: $a_o = 3.970$A, $b_o = 3.544$A, and $c_o = 12.84$A

Space group: D_{2h}^{25} — $Immm$; n.o. = not observed.

d obs., A	d calc., A	I obs.	I calc.	hkl
6.44	6.42	10.2	22.4	002
n.o.	3.793	n.o.	0.1	101
3.419	3.416	3.2	3.3	011
3.217	3.210	2.6	2.4	004
2.912	2.911	64.0	67.6	103
2.730	2.730	73.9	76.2	013
2.643	2.644	100.0	98.9	110
2.447	2.445	1.8	2.2	112
2.141	2.156	33.3	10.2	105
	2.140		22.2	006
2.079	2.079	12.8	14.0	015
2.041	2.041	6.4	6.4	114
1.985	1.985	25.8	25.2	200
1.896	1.896	1.5	1.9	202
1.772	1.772	18.9	17.2	020
1.711	1.716	1.7	0.6	211
	1.708		1.4	022
	1.688		0.6	204
1.663	1.665	35.2	0.9	107
	1.663		30.3	116
n.o.	1.629	n.o.	0.2	017
1.605	1.605	36.8	30.0	213
	1.605		5.1	008
n.o.	1.551	n.o.	0.4	024
1.513	1.514	20.4	19.7	123
1.455	1.455	14.3	11.8	206
1.436	1.436	6.2	7.8	215
1.368	1.372	21.8	9.2	118
	1.369		4.6	125
	1.365		9.3	026
1.342	1.343	2.9	2.2	109
1.322	1.323	14.7	3.0	019
	1.322		12.2	220

Table III. (Continued)

Orthorhombic: $a_o = 3.970$A, $b_o = 3.544$A, and $c_o = 12.84$A

Space group: D_{2h}^{25} — $Immm$; n.o. = not observed.

d obs., A	d calc., A	I obs.	I calc.	hkl
n.o.	1.295	n.o.	1.1	222
n.o.	1.284	n.o.	0.2	0·0·10
1.264	1.264	6.7	5.4	303
	1.259		0.2	217
1.241	1.248	11.5	4.3	208
	1.240		8.3	310
n.o.	1.222	n.o.	0.4	224
n.o.	1.217	n.o.	0.4	312
n.o.	1.213	n.o.	0.6	127
1.189	1.190	3.0	3.8	028
1.176	1.176	1.0	0.1	031
	1.176		1.5	305
1.155	1.156	2.3	0.7	314
	1.155		1.1	1·1·10
1.136	1.139	10.1	4.9	033
	1.132		6.4	130
1.124	1.125	17.7	10.0	226
	1.120		4.3	1·0·11
	1.115		0.3	132
1.108	1.109	4.6	4.9	0·1·11
1.101	1.101	4.2	3.5	219

002 reflection was excluded, since this low-angle reflection did not receive the full intensity of the x-ray beam.

With the assumed atomic coordinates, divalent palladium is surrounded by four coplanar (rectangular) oxygens, two at 1.985A and two at 2.054A. The strontium atoms are surrounded by seven nearest oxygen neighbors, four at 2.67A, two at 2.57A, and one at 2.50A. This leads to an average Sr–O separation of 2.62A, which is in very good agreement with the sum of the ionic radii, 2.61A (17). The above distances are based on the two approximate z parameters ($z = 0.355$ for Sr and $z = 0.16$ for O in $4i$) which are believed to be accurate within ± 0.01A. These two variable z-parameters were chosen on the basis of a few trial calculations. We have not bothered to do a refinement with the powder data. Any future refinement is best carried out with single-crystal data, since single crystals are easily obtainable.

The structure can be considered as an orthorhombically distorted K_2NiF_4 type, from which $\frac{1}{4}$ of the oxygens have been removed in an ordered way. If these missing oxygens are inserted—2 O in $(2d)$, $\frac{1}{2}0\frac{1}{2}$ + b.c. in space group $Immm$—the stoichiometry Sr_2PdO_4 is obtained. As has already been indicated, our attempts to determine the oxygen content in Sr_2PdO_3 have failed to yield a value consistent with the assumed formula. Somewhat higher oxygen contents are obtained, up to the $Sr_2PdO_{3.6}$ stoichiometry. After considering the possibility that the $(2d)$ positions in the structure are occupied by significant amounts of oxygen, we rejected this hypothesis for two reasons:

(i) With increasing amounts of oxygen in $(2d)$, the intensity agreement becomes significantly poorer, e.g., the 101 reflection becomes comparable in intensity to the 011 and the 103 reflection becomes more intense than the 013, in contradiction to our observations.

(ii) Abnormally short Pd–O distances (1.772A) would be formed, at least 0.2A shorter than the expected bond length.

We believe that the high oxygen content may possibly be caused by sorption of H_2O or CO_2 prior to analysis since SrO is highly hygroscopic and absorbs CO_2 from the air.

Addendum. While this article was being refereed, a paper appeared describing the crystal structure of Sr_2CuO_3 ([20]). It is apparent that Sr_2PdO_3 and Sr_2CuO_3 are isotypic, although the German authors ([20]) have chosen for Sr_2CuO_3 a different origin with the $a,b,$ and c-axes interchanged, compared with those selected for Sr_2PdO_3. This difference in choice can be attributed to the fact that we have derived the structure of Sr_2PdO_3 from the K_2NiF_4 type and therefore have chosen similar settings. The structure of Sr_2CuO_3 was derived from Patterson and Fourier syntheses without drawing any analogy to the K_2NiF_4 structure.

$SrPd_3O_4$, $CaPd_3O_4$, and $CdPd_3O_4$. The three new phases, $SrPd_3O_4$, $CaPd_3O_4$, and $CdPd_3O_4$, can be prepared easily by heating the reagent grade mixtures at 900°C and 200 atm of oxygen. However, high oxygen pressures are not required, and all three phases can be made by heating the starting mixtures in air. If the synthesis is carried out in air, the products are generally not pure and contain significant amounts of end-component oxides and/or palladium metal. To help eliminate these impurities, the following points should be noted:

(i) The starting mixture should contain some excess MeO oxide or hydroxide above the stoichiometric 1:3 ratio. This results in the formation of some excess Sr_2PdO_3, CaO, or CdO. All three of these impurities can be dissolved away with dilute nitric acid, leaving the insoluble $MePd_3O_4$ behind.

(ii) Palladium metal impurities can be removed with aqua regia. However, $SrPd_3O_4$, $CaPd_3O_4$, and $CdPd_3O_4$ will dissolve slowly in aqua regia, at least to some extent. Thus, any treatment with aqua regia should not be prolonged.

(iii) Regrinding, remixing, and reheating the compressed pellets may help reduce the impurities.

(iv) The synthesis of $CdPd_3O_4$ in air is best carried out at 800°–820°C, while for $SrPd_3O_4$ and $CaPd_3O_4$ ~950°C is recommended. Lower

Table IV. X-Ray Diffraction Powder Data for $SrPd_3O_4$

Primitive Cubic: a_o = 5.826A. S.G.: o_h^3-$Pm3n$ ($Na_xPt_3O_4$ structure)

(n.o. = not observed)

d obs., A	d calc., A	I obs.	I calc.	hkl
4.13	4.12	1.4	1.7	110
2.913	2.913	12.0	12.2	200
2.606	2.605	100.0	105.6	210
2.379	2.378	61.6	65.8	211
2.060	2.060	1.0	1.0	220
1.842	1.842	0.4	0.4	310
1.682	1.682	17.2	17.1	222
1.616	1.616	27.1	24.1	320
1.557	1.557	36.4	34.0	321
1.456	1.456	21.9	19.2	400
1.374	{1.373, 1.373}	0.3	{0.1, 0.2}	330, 411
1.303	1.303	5.6	5.2	420
1.271	1.271	22.9	21.4	421
1.242	1.242	8.3	7.9	332
n.o.	1.189	n.o.	0.1	422
1.142	{1.143, 1.143}	0.3	{0.2, 0.1}	431, 510
1.0820	{1.0819, 1.0819}	21.1	{7.0, 13.8}	520, 432
1.0637	1.0637	10.8	10.5	521
1.0302	1.0299	13.7	13.7	440
n.o.	{0.9991, 0.9991}	n.o.	{0.1, 0.1}	530, 433
0.9710	{0.9710, 0.9710}	3.9	{0.7, 2.9}	600, 442
0.9576	0.9578	6.2	5.9	610
0.9451	{0.9451, 0.9451}	14.8	{4.6, 9.3}	611, 532
n.o.	0.9212	n.o.	0.0	620
n.o.	0.8990	n.o.	0.2	541
0.8782	0.8783	11.1	11.1	622
0.8685	{0.8685, 0.8685}	19.9	{6.8, 13.4}	630, 542
0.8589	0.8590	10.1	10.9	631
0.8409	0.8409	10.2	10.9	444

temperatures give lower conversion rates, while significantly higher temperatures result in the decomposition of the products.

Typical analyses of the products yielded the following formulas: $Sr_{1.01}Pd_{3.00}O_{4.22}$, $Ca_{0.99}Pd_{3.00}O_{4.05}$, and $Cd_{0.95}Pd_{3.00}O_{4.03}$. The strontium compound gives higher than expected oxygen contents, especially when the samples are prepared at high oxygen pressures.

The x-ray powder data for SrPd$_3$O$_4$ are given in Table IV. CaPd$_3$O$_4$ and CdPd$_3$O$_4$ give very similar powder patterns. All three patterns can be indexed on the basis of a primitive cubic unit cell with cell constants of 5.826, 5.747, and 5.742A for SrPd$_3$O$_4$, CaPd$_3$O$_4$, and CdPd$_3$O$_4$, respectively. Intensity calculations were carried out in a similar manner as for Sr$_2$PdO$_3$, assuming the Na$_x$Pt$_3$O$_4$ structure (22, 23). The atoms were placed as follows in space group o$_h^3$-$Pm3n$:

2 Sr (Ca or Cd) in (2a) : $(000, \frac{1}{2}\frac{1}{2}\frac{1}{2})$

6 Pd in (6c) : $\pm (\frac{1}{4}0\frac{1}{2}, \circlearrowleft)$

8 O in (8e) : $\pm (\frac{111}{444}, \frac{133}{444} \circlearrowleft)$

Table V. Structural Data for Compounds with the Na$_x$Pt$_3$O$_4$ Structure

Cubic, S.G.: o$_h^3$ — $Pm3n$

Interatomic Distances, A^a

Composition	a$_0$, A	Pt–O or Pd–O, square planar	Me–O, 8-fold, cubic	Shortest, Pt–Pt or Pd–Pd	Reference
Pt$_3$O$_4$	5.585	1.974	2.418	2.792	(7)
Mg$_x$Pt$_3$O$_4$	5.621	1.987	2.434	2.810	(11)
Na$_x$Pt$_3$O$_4$	5.69	2.01	2.46	2.84	(22, 23)
Na$_x$Pd$_3$O$_4$	5.64	1.99	2.44	2.82	(16)
CdPd$_3$O$_4$	5.742	2.030	2.486	2.871	Present work
CaPd$_3$O$_4$	5.747	2.032	2.488	2.873	Present work
SrPd$_3$O$_4$	5.826	2.059	2.523	2.913	Present work

^{a} The interatomic distances for CdPd$_3$O$_4$, CaPd$_3$O$_4$, and SrPd$_3$O$_4$ are believed to be accurate to within ± 0.002A. This small error is a function only of the accuracy of the a$_0$ cell edge measurements, since there are no variable positional parameters in the Na$_x$Pt$_3$O$_4$ structure.

Table IV shows that the intensity agreement is quite good. The intensity discrepancy factors $R = 100 \ (\Sigma |I_o - I_c|)/\Sigma(I_o)$ were calculated as 5.8, 7.8, and 8.4 for SrPd$_3$O$_4$, CaPd$_3$O$_4$, and CdPd$_3$O$_4$, respectively. In Table V, the interatomic distances for these compounds are compared with the distances found for other Na$_x$Pt$_3$O$_4$-type phases. The square planar Pd–O bond lengths and the cubic (eight-coordinated) Sr–O, Ca–O, and Cd–O distances are in good agreement with the expected values. Square planar coordination is common for divalent palladium compounds, and the eight-fold (cubic) environment for Sr^{2+}, Ca^{2+}, and Cd^{2+} is expected for these ions. The relatively short Pd–Pd distances may indicate the presence of some inter-metallic bonding. In these compounds, the palladium atoms are apparently completely ordered in the (6c) sites. We have found no x-ray evidence for any noticeable disorder between the two different cations in $CaPd_3O_4$. Because of the strong preference of Pd^{2+} for square planar coordination, it can be assumed that no such disorder exists in $SrPd_3O_4$ and $CdPd_3O_4$, although this could not be proved easily by x-ray methods since the atomic scattering factors for Sr, Cd, and Pd are very similar.

Mg_2PdO_4. When $Mg(OH)_2$–Pd black mixtures are heated at high oxygen pressures—e.g., at 900°C and 200 atm of oxygen—a poorly crystallized spinel phase is formed with $a_o = 8.501A$. We have not yet succeeded in preparing this spinel in a pure state, since this phase is always accompanied by some unreacted MgO and PdO. From crystal–chemical considerations and from the x-ray intensities, we have tentatively assigned the stoichiometry Mg_2PdO_4 to this phase. Mg_2PdO_4 is apparently an inverse spinel with Pd^{4+} occupying octahedral sites. The unit cell constant for Mg_2PdO_4, $a_o = 8.501A$, is slightly smaller than the corresponding value for Mg_2PtO_4, $a_o = 8.521A$ (10), owing to the slightly smaller ionic radius for Pd^{4+} (17).

Acknowledgment

We thank W. A. Jester for the neutron activation analyses. This work was supported by the Advanced Research Projects Agency, Contract No. DA-49-083 OSA-3140, and by the U. S. Army Electronics Command, Contract DA28-043 AMC-01304 (E).

Literature Cited

(1) Guiot, J. M., *J. Appl. Phys.* **1968**, 39, 3509.
(2) Hoekstra, Henry R., Siegel, Stanley, Gallagher, Francis X., ADVAN. CHEM. SER. **1970**, 98, 39.
(3) McDaniel, C. M., Schneider, S. J., *J. Res. Natl. Bur. Std.* **1968**, 72A, 27.
(4) Moore, W. J., Pauling, L, *J. Am. Chem. Soc.* **1941**, 63, 1392.
(5) Muller, O., Roy, R., *Am. Ceram. Soc. Bull.* **1967**, 46, 881.
(6) Muller, O., Roy, R., *J. Inorg. Nucl. Chem.* **1969**, 31, 2966.
(7) Muller, O., Roy, R., *J. Less-Common Metals* **1968**, 16, 129.
(8) *Ibid.*, **1969**, 19, 209.
(9) *Ibid.*, **1970**, 20, 161.
(10) Muller, O., Roy, R., *Mater. Res. Bull.* **1969**, 4, 39.
(11) Muller, O., Roy, R., unpublished data.
(12) Randall, J. J., Katz, L., Ward, R., *J. Am. Chem. Soc.* **1957**, 79, 266.
(13) Randall, J. J., Ward, R., *J. Am. Chem. Soc.* **1959**, 81, 2629.
(14) Sabrowsky, H., Hoppe, R., *Naturwissenschaften* **1966**, 53, 501.
(15) Scheer, J. J., Dissertation, Leiden, Netherlands, 1956.
(16) Scheer, J. J., van Arkel, A. E., Heyding, R. D., *Can. J. Chem.* **1955**, 33, 683.

(17) Shannon, R. D., Prewitt, C. T., *Acta Cryst.* **1969,** B25, 925.
(18) Shannon, R. D., Rogers, D. B., Prewitt, C. T., in press.
(19) Sleight, A. W., *Mater. Res. Bull.* **1968,** 3, 699.
(20) Teske, C. L., Muller-Buschbaum, H., *Z. Anorg. Allgem. Chem.* **1969,** 371, 325.
(21) Waser, J., Levy, H. A., Peterson, S. W., *Acta Cryst.* **1953,** 6, 661.
(22) Waser, J., McClanahan, E. D., *J. Chem. Phys.* **1951,** 19, 199.
(23) *Ibid.,* **1951,** 19, 413.

RECEIVED February 2, 1970.

4

Reaction of Platinum Dioxide with Some Metal Oxides

HENRY R. HOEKSTRA, STANLEY SIEGEL, and
FRANCIS X. GALLAGHER

Argonne National Laboratory, Argonne, Ill. 60439

> *The structure of platinum dioxide and its reactions with some di, tri, and tetravalent metal oxides have been investigated. Ternary platinum oxides were synthesized at high pressure (40 kilobars) and temperature (to 1600°C). Properties of the systems were studied by x-ray, thermal analysis, and infrared methods. Complete miscibility is observed in most PtO_2–rutile-type oxide systems, but no miscibility or compound formation is found with fluorite dioxides. Lead dioxide reacts with PtO_2 to form cubic $Pb_2Pt_2O_7$. Several corundum-type sesquioxides exhibit measurable solubility in PtO_2. Two series of compounds are formed with metal monoxides: M_2PtO_4 (where M is Mg, Zn, Cd) and MPt_3O_6 (where M is Mg, Co, Ni, Cu, Zn, Cd, and Hg).*

The literature on anhydrous binary or ternary platinum oxides is relatively limited. Muller and Roy (6) have reviewed the simple platinum oxides and were able to confirm the existence of only two oxides, Pt_3O_4 and PtO_2, with the dioxide existing in two crystal modifications. Shannon (13) has discussed the preparation and properties of orthorhombic PtO_2. Ternary oxides of platinum include several with the alkali metals ($Na_xPt_3O_4$, Na_2PtO_3, Li_2PtO_3) (12) and with the alkaline earths [$Ba_3Pt_2O_7$ (18), $Sr_3Pt_2O_7$, and Sr_4PtO_6 (10)]. More recent work has been reported on $Tl_2Pt_2O_7$ (4) and the rare earth–platinum pyrochlores (3, 17), as well as the spinels Zn_2PtO_4 and Mg_2PtO_4 (5).

The investigations of platinum pyrochlores have demonstrated the effectiveness of high pressure techniques in the synthesis of anhydrous oxides when one or both reactants have limited thermal stability. The bulk of the work reported here represents a continuation of an exploration of metal oxide–platinum oxide systems at high pressure.

Equipment and Procedure

The high pressure apparatus used in this work is a tetrahedral anvil apparatus. Details of the equipment, the operating procedure, and the sample assembly have been described previously (2) and will not be repeated here. Briefly, the powdered oxide samples are wrapped in platinum or in gold foil to isolate them, as much as possible, from reaction with materials other than the reactant mixture. We shall see that on occasion the oxide samples will react with the platinum foil container to give lower valence platinum oxides. As a rule, three oxide samples, which have been wrapped in foil and pressed into pellets, are loaded into the sample cavity (~ 1 cc volume) of a pyrophyllite tetrahedron for each high pressure run.

In a typical experiment, the tetrahedron with its three samples is compressed to 40 kb pressure; the sample cavity is heated to 1200°C and maintained at that temperature for an hour before being quenched to room temperature. Finally, pressure is released slowly. Although the bulk of our experiments have been run at 40 kb pressure, the range in pressure applied has varied from 20 to 60 kb; experimental temperatures varied from 500° to 1600°C.

We find that under these experimental conditions loss of an appreciable amount of oxygen from the sample pellet occurs very rarely, even though the oxides are heated to several hundred degrees above their decomposition temperature at atmospheric pressure. Exceptions to this rule occur with the formation of any stable oxide phase which does not utilize all of the oxygen present in the initial reactant mixture. This "excess" oxygen is lost rather rapidly from the sample enclosure.

Our investigation of the properties of the platinum oxides includes x-ray, infrared, and thermal analysis. Powder x-ray data were obtained with a Philips 114.59 mm camera, using either Ni-filtered Cu radiation or V-filtered Cr radiation. Cell parameters were obtained using a least squares refinement, and are based on Cu $K\alpha = 1.5418$ A and Cr $K\alpha = 2.2909$. The *dta–tga* results were obtained on a Mettler recording thermoanalyzer over the temperature interval 25°–1400°C, with a heating rate of 6°/minute and a sensitivity of 100 μV. Infrared spectra to 200 cm^{-1} were taken with a Beckman IR-12 spectrophotometer. Nujol mulls were spread on polyethylene disks for the low frequency portion of the spectrum and on KBr plates for the frequencies above 400 cm^{-1}. Platinum dioxide was prepared by the reaction of K_2PtCl_4 or K_2PtCl_6 (99.9% purity) with molten KNO_3 at 400°C. The dioxide was recovered by solution of the KNO_3 in water and purified by heating in aqua regia for several hours. Spectrochemical analysis of the product showed less than 0.2% of alkali metals, and no water or hydroxyl ion was detect-

Table I. X-ray Powder Data for α-PtO$_2$

Hexagonal a = 3.100(2)   c = 4.161(3) Å

hkl	d	I_{obsd}	I^a_{calcd}
001	4.107	60	100
100	2.664	100	100
101	2.241	90	170
002	2.065	10	15
102	1.637	60	100
110	1.543	90	50
111	1.448	70	70
003	1.381	1	10
200	1.337	45	25
201	1.2734	50	50
112	1.2388	30	40
103	1.2287	15	50
202	1.1255	35	50
113	1.0309	5	40
210	1.0126	55	40
211	0.9841	55	90
203	0.9628	20	45
212	0.9111	60	115
300	0.8945	35	30
301	0.8752	40	60
114	0.8629	20	80
302	0.8218	100	75

^{a} Based on CdI$_2$–type structure.

able by infrared analysis. Metal oxides used in the reactions with PtO$_2$ were usually reagent grade chemicals; exceptions included oxides such as FeO, CrO$_2$, MnO, Mn$_2$O$_3$, VO$_2$, and MoO$_2$, which were prepared by accepted procedures from reagent grade starting materials. Purity of these products was confirmed by x-ray, thermal, and chemical analysis.

Results

α-PtO$_2$. Several proposals concerning the structure of the ambient pressure form of PtO$_2$ have been reviewed by Muller and Roy (6). Difficulties arise principally from the poor crystallinity which characterizes this phase, regardless of the heating time employed in its preparation. Disorder is particularly troublesome in the c-direction of the hexagonal structure since hkl lines with $1 \neq 0$ become increasingly weak and diffuse with increase in l.

Our experience with the synthesis of α-PtO$_2$ confirms previous reports on the characteristic disordered structure of the oxide. Vaughn

Leigh, one of our student aides, noted that the thin film of PtO_2 which forms on the surface of the molten KNO_3 shows much less disorder and gives a much improved powder pattern over the bulk dioxide in the nitrate melt. Table I gives the x-ray powder data obtained from such a PtO_2 surface layer. The calculated hexagonal cell parameters are $a = 3.100(2)$ and $c = 4.161(3)$ A; the x-ray density is 10.90 grams/cm^3.

We prefer the layered CdI_2-type structure (space group $P\bar{3}m1 - D_{3d}{}^3$) for α-PtO_2. Even though the intensity agreement is only fair in our "ordered" phase, it is much superior to that of the bulk material, and would be expected to show further improvement as ordering is increased in the c-direction. If one assigns oxygen atoms to the $(1/3, 2/3, z)$ and $(2/3, 1/3, \bar{z})$ positions with $z = 0.25$, the calculated platinum–oxygen bond length is 2.07 A. This is 0.08 A longer than is observed in β-PtO_2 (16), or derived from Shannon-Prewitt radii (14) (note: all radii used in this discussion are taken from the Shannon-Prewitt tables), but is reasonable for the more open structure of the ambient pressure phase. The oxygen–oxygen distances are 2.74 A between layers and 3.10 A within each layer.

Further support for the CdI_2-type structure in α-PtO_2 is afforded by the recent synthesis of two additional forms of PtO_2 in sealed tube experiments *via* the reaction

$$K_2PtCl_4 + KNO_3 \rightarrow PtO_2 + 3KCl + NOCl \tag{1}$$

Heating times at 425°C were 5 to 25 days. One modification appears to have a unit cell 3 times the α-PtO_2 cell in the c-direction, with $a = 3.11$

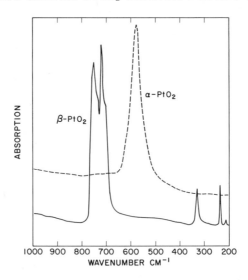

Figure 1. Infrared spectra of PtO_2 phases

(1) A and $c = 12.60(6)$ A. This form seems to be related to the CdI_2-type III reported by Pinsker (9). The second modification can be indexed on the basis of CdI_2-type II with $a = 3.10(1)$ A and $c = 8.32(6)$ A; the α-PtO_2 cell is doubled in the c-direction. The space group for this form is C_{6v}^4-$P6_3mc$. With coordinate values based upon Cd and I positions, the Pt–O bond lengths are about 2.07 A in both modifications. Thus, the simple and multiple hexagonal cells are known for both CdI_2 and ambient pressure PtO_2.

β-PtO_2. We have not succeeded in preparing samples of the high pressure form of PtO_2 suitable for single crystal studies. A structural analysis of orthorhombic β-PtO_2 has been carried out, however, based upon the intensities of diffraction lines from powder samples (16). The cell parameters are $a = 4.488(3)$, $b = 4.533(3)$, $c = 3.138(2)$ A. The intensity data confirm the tentative $CaCl_2$-type structure with oxygen coordinates $x = 0.281$ and $y = 0.348$ to give two long Pt–O bonds of 2.02 A and four shorter bonds of 1.98 A.

The infrared spectra of α and β-PtO_2 are shown in Figure 1. The single absorption maximum in α-PtO_2 is analogous to that observed with CdI_2 (11). The strong peaks of β-PtO_2 are assigned to Pt–O asymmetric stretching modes, and the weak lower frequency bands are attributed to bending modes in the PtO_2 lattice. The higher frequencies observed in the β-PtO_2 stretching vibrations (750, 720 cm^{-1}) over that in α-PtO_2 (580 cm^{-1}) are consistent with the ~ 0.1 A shorter platinum–oxygen bond lengths in the β-form. The α-PtO_2 spectrum is inconsistent with the structure proposed by Busch et al. (1) since the extremely short (1.47 A) Pt–O distance in the Busch structure would require a much higher stretching frequency than we observe for this phase.

Both modifications of PtO_2 are insoluble in hot concentrated HNO_3, H_2SO_4, and aqua regia. α-PtO_2, however, dissolves slowly in hot concentrated HCl, more readily in HBr; β-PtO_2 is soluble only in HBr.

Thermal analysis of α-PtO_2 indicates that oxygen evolution begins at 575°C. Decomposition to the metal occurs in a single step, and is complete at $\sim$600°C. The β-form is somewhat more stable, with decomposition beginning at 600°C. The dta–tga trace frequently gives an indication of a two-step process, but all attempts to isolate or identify an intermediate product have been unsuccessful. All x-ray powder patterns of partially decomposed samples showed only two phases: PtO_2 and Pt.

$MO_2 + PtO_2$

The reaction of platinum dioxide with other metal dioxides can be separated into three groups—i.e., compounds with metal–oxygen coordination numbers of 4:2, 6:3, and 8:4. We have investigated only a single

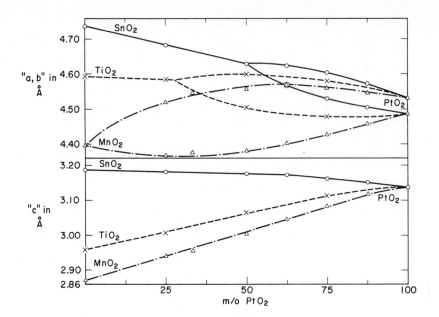

Figure 2. Cell dimensions of $MO_2 \cdot PtO_2$ solid solutions

4:2 coordinated oxide, SiO_2 (GeO_2 also exists in the α-quartz structure but not at high pressure). We find no evidence for compound formation or solid solution in the SiO_2–PtO_2 system at 40 kb and 1200°C. The silica portion of the sample is converted to coesite and the β-PtO_2 diffraction pattern shows no measurable shift in line positions.

Mixtures of PtO_2 with 10 oxides having the rutile or a distorted rutile structure have been investigated. Two of the 10, MoO_2, and WO_2, reduce PtO_2 to the metal and are themselves oxidized to the trioxides. No evidence for further reaction between MoO_3 or WO_3 and excess PtO_2 could be detected from an x-ray investigation of the products.

Seven 6:3 coordinated dioxides (TiO_2, VO_2, CrO_2, MnO_2, GeO_2, RuO_2, and SnO_2) show rather similar reaction towards PtO_2, in that extensive solid solution occurs at 40 kb pressure (750° to 1200°C) without evidence of compound formation. Some of the systems were investigated at 25 m/o intervals, others at 12.5 m/o intervals. Incomplete miscibility was detected in only one of the seven, the CrO_2–PtO_2 system. A single phase region is found from ~50 to 100 m/o PtO_2, but PtO_2 shows little solubility in CrO_2. It is possible that more detailed studies will disclose a similar, but narrower, two-phase region between 0 and 25 m/o PtO_2 in the MnO_2–PtO_2 system.

In each of the seven systems studied, the addition of MO_2 to PtO_2 initially causes an increase in the b/a ratio of the orthorhombic PtO_2 phase.

Table II. MO_2–PtO_2 Cell Dimensions in Angstroms[a]

Composition	a	b	c
GeO_2	4.390	—	2.859
$7GeO_2 \cdot PtO_2$	4.410(5)	—	2.891(3)
$3GeO_2 \cdot PtO_2$	4.432	—	2.932
$5GeO_2 \cdot 3PtO_2$	4.444	—	2.969
$GeO_2 \cdot PtO_2$	4.465	—	3.009
$3GeO_2 \cdot 5PtO_2$	4.437	4.514	3.033
$GeO_2 \cdot 3PtO_2$	4.437	4.546	3.069
$GeO_2 \cdot 7PtO_2$	4.448	4.549	3.102
PtO_2	4.488	4.533	3.138
SnO_2	4.737	—	3.186
$3SnO_2 \cdot PtO_2$	4.682	—	3.182
$SnO_2 \cdot PtO_2$	4.630	—	3.176
$3SnO_2 \cdot 5PtO_2$	4.571	4.625	3.172
$SnO_2 \cdot 3PtO_2$	4.531	4.603	3.162
$SnO_2 \cdot 7PtO_2$	4.506	4.574	3.151
VO_2[b]	4.517	—	2.872
$3VO_2 \cdot PtO_2$	Tetragonal not measured		
$VO_2 \cdot PtO_2$	4.503(5)	—	3.051(3)
$VO_2 \cdot 3PtO_2$	4.455	4.553	3.092
TiO_2	4.594	—	2.958
$3TiO_2 \cdot PtO_2$	4.585(5)	—	3.007(5)
$TiO_2 \cdot PtO_2$	4.506(5)	4.600(5)	3.065(5)
$TiO_2 \cdot 3PtO_2$	4.479	4.580	3.113
RuO_2	4.491	—	3.105
$3RuO_2 \cdot PtO_2$	4.498	—	3.119
$RuO_2 \cdot PtO_2$	4.475	4.539	3.129
$RuO_2 \cdot 3PtO_2$	4.469	4.559	3.137
MnO_2	4.396	—	2.871
$3MnO_2 \cdot PtO_2$	4.368(5)	4.519(5)	2.941(5)
$2MnO_2 \cdot PtO_2$	4.377(5)	4.547(5)	2.956(3)
$MnO_2 \cdot PtO_2$	4.382(5)	4.557(5)	3.006(3)
$3MnO_2 \cdot 5PtO_2$	4.402	4.571	3.047
$MnO_2 \cdot 3PtO_2$	4.428	4.561	3.083
$MnO_2 \cdot 7PtO_2$	4.459	4.549	3.188
CrO_2	4.422	—	2.918
$3CrO_2 \cdot PtO_2$		2 phases	
$CrO_2 \cdot PtO_2$	4.438(5)	4.564(5)	3.109(3)
$CrO_2 \cdot 3PtO_2$	4.465	4.554	3.157

[a] Error limits are ± 0.002 A except as indicated, e.g., (5).
[b] Based on rutile-type cell.

Further MO_2 additions eventually reverse this trend as the b/a ratio decreases to unity and the solid acquires the tetragonal rutile structure. Figure 2 illustrates three of the systems: (a) SnO_2–PtO_2 exhibits tetragonal symmetry over ~50% of the composition range; GeO_2 and VO_2 show similar symmetry patterns with PtO_2. (b) Tetragonal symmetry is found over ~25% of the TiO_2–PtO_2 and RuO_2–PtO_2 systems, and (c) little if any tetragonal range can be observed in PtO_2 mixtures with CrO_2 or MnO_2. Table II lists the cell parameters measured for the MO_2–PtO_2 mixtures investigated. The symmetry relationships in MO_2–PtO_2 systems are different from those observed by Magneli and coworkers (19) in mixtures of other distorted rutile-type structures. In these systems, tetragonal structures were observed over the major portion of the composition range, whereas the maximum extent of tetragonal symmetry is only ~50% in the seven Mo_2–PtO_2 under investigation here.

Thermal analysis of MO_2–PtO_2 solid solutions reveal no marked stabilization effect in those mixtures containing dioxides of limited thermal stability, such as CrO_2 or MnO_2. The stable more refractory dioxides, however, increase the stability of the PtO_2 component of the solid solution. Thus, only 1/3 of the oxygen associated with platinum in the $75SnO_2 \cdot 25PtO_2$ mixture is evolved as the solid solution is heated to 1250°C at 6°/minute.

PbO_2. Lead dioxide crystallizes in a rutile structure and could be expected to show at least partial solubility in β-PtO_2. Under pressure, however, PbO_2 converts to an orthorhombic modification which is virtually insoluble in PtO_2, and represents a notable exception to the extensive solid solution range observed in related systems from GeO_2–PtO_2 to SnO_2–PtO_2.

Of further interest in the lead–platinum–oxygen system is the observation that a 1:1 mixture of the dioxides will react at 40 kb pressure and at temperatures above 750°C to form a phase indexible as a face-centered cubic structure with $a = 5.15(1)$ A. Very little solid solution occurs on either side of the 1:1 metal ratio since deviation from this ratio gives a diphasic mixture of the cubic phase (with no discernible cell parameter change) and either PtO_2 or PbO_2. Thermal analysis of the cubic phase has established that decomposition to PbO and Pt occurs at 740°C, and that 2.5 atoms of oxygen are evolved for each lead or platinum atom. These data strongly suggest that the new compound is $Pb_2Pt_2O_7$, having a cubic pyrochlore-type structure with $a = 10.30$ A, twice the apparent unit cell dimension, rather than a disordered fluorite structure. This interpretation requires the presence of trivalent lead in the ternary oxide. A comparison of its unit cell dimension with that of the rare earth–platinum pyrochlores indicates that the presumed Pb(III) ion has a radius of 1.08 A for a coordination number of 8. This is reasonable when compared with its trivalent neighbors, Tl(III) at 1.00 A and Bi(III) at 1.11 A.

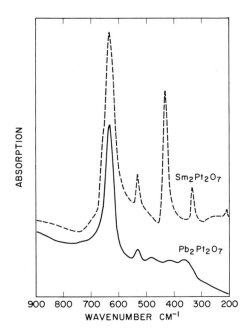

Figure 3. Infrared spectra of $Pb_2Pt_2O_7$ and $Sm_2Pt_2O_7$

The infrared spectrum of $Pb_2Pt_2O_7$ shows a single strong absorption peak at 633 cm^{-1} which is assigned to the asymmetric stretching mode in the octahedral oxygen configuration about platinum. The comparable peak in $Sm_2Pt_2O_7$ ($a = 10.31$ A) is found (Figure 3) at 635 cm^{-1}. The remainder of the $Pb_2Pt_2O_7$ spectrum consists of three weak maxima at 530, 480, and 350 cm^{-1}. The peak locations correspond reasonably well with maxima in the $Sm_2Pt_2O_7$ spectrum, but the peak intensities are much weaker. The reason for this difference is not understood at present.

Fluorite Structures. The reaction of PtO_2 with several 8:4 coordinated metal dioxides has been investigated. Neither CeO_2 nor ThO_2 gives any indication of reaction with PtO_2 at 40 kb and 1200°C, either by way of solid solution or compound formation. Uranium dioxide is oxidized by PtO_2 under these conditions to the high pressure form of UO_3 (15), with no indication of reaction between PtO_2 and UO_3. Zirconium dioxide, a distorted fluorite phase, does not react with PtO_2 under these experimental conditions.

$M_2O_3 + PtO_2$

Rare Earth Sesquioxides. Oxides of the larger trivalent metal ions react (3, 17) with PtO_2 at high pressure to form a pyrochlore series,

$A_2Pt_2O_7$, similar to those reported for tin, ruthenium, titanium, and several other tetravalent ions. Trivalent ions which form cubic platinum pyrochlores range from Sc(III) at 0.87 A to Pr(III) at 1.14 A. Distorted pyrochlore structures are formed by lanthanum (1.18 A) and by bismuth (1.11 A). Platinum dioxide oxidizes Sb_2O_3 to Sb_2O_4 at high pressure. The infrared spectra and thermal stability of the rare earth platinates have been reported previously and will not be repeated here, except to point out the rather remarkable thermal stability of these compounds; decomposition to the rare earth sesquioxide and platinum requires temperatures in excess of 1200°C.

Smaller Trivalent Ions. Mixtures of aluminum, vanadium, chromium, manganese, iron, and gallium sesquioxide with PtO_2 were investigated for possible compound formation—none was found. The trivalent ions which have stable tetravalent oxidation states—*i.e.*, vanadium, manganese, and chromium—are oxidized by PtO_2 to form MO_2–PtO_2 solid solutions which have been discussed previously. Aluminum, iron, and gallium sesquioxides show a partial solubility in the β-PtO_2 phase to form oxygen-deficient solid solutions. Figure 4 illustrates the change in PtO_2 unit cell volume as a function of mole fraction Pt. The Al_2O_3 solubility is less than 3 m/o (the lowest concentration investigated), but the Fe_2O_3 solubility is approximately 15 m/o. The Ga_2O_3 solubility is difficult to determine from x-ray data since very little change in cell parameters is

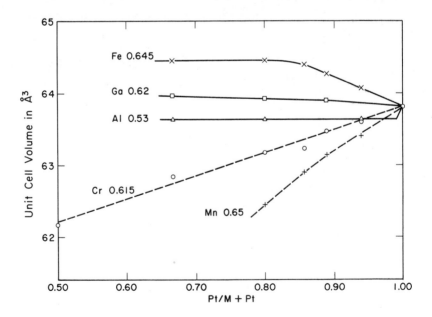

Figure 4. Volume change in PtO_2 unit cell with M_2O_3 addition

observed, but the solubility is estimated to be 10 m/o. The solid solutions are assumed to be oxygen-deficient rather than acquiring interstitial cations since Fe(III) with an ionic radius (0.645 A) greater than Pt(IV) (0.63 A) brings about an increase in cell volume, while Ga(III) (0.62 A) produces virtually no change, and Al(III) at 0.53 causes a sharp decrease in cell volume. The anomalous effect observed in the Cr_2O_3 and Mn_2O_3 mixtures is attributed to a combined oxidation plus solid solution reaction. Our x-ray data indicate that complete oxidation to CrO_2 does not occur since cell parameters obtained from CrO_2 + PtO_2 mixtures do not coincide with those derived from Cr_2O_3 + PtO_2 reactants (after correcting for PtO_2 used in oxidation of chromium). Oxidation of Mn_2O_3 to MnO_2 is complete, since comparable cell dimensions are obtained from the two reactant mixtures.

MO + PtO_2

The present discussion will not include results on reactions of PtO_2 with CaO, SrO, or BaO. This portion of our investigation has not been completed and will be reported later. Several of the remaining systems need be mentioned only briefly: PdO shows no evidence of reaction with PtO_2 at high pressure, and the readily oxidizable monoxides of manganese, iron, and tin reduce PtO_2 to the metal.

M_2PtO_4 Compounds. Muller and Roy (5) have described the synthesis of spinels in the ZnO–PtO_2 and MgO–PtO_2 systems. We have prepared these compounds at high pressure; the unit cell dimensions are 8.550(2) A for Zn_2PtO_4 and 8.521(2) A for Mg_2PtO_4, in excellent agreement with Muller and Roy. An experiment at 1550°C has produced bright yellow octahedra of the zinc spinel which are suitable for single crystal studies. Thermal analysis of this crystalline material indicates that endothermic decomposition to ZnO and Pt occurs at 830°C.

The infrared spectra of Zn_2PtO_4 (Figure 5) and Mg_2PtO_4 are characteristic of spinels with three strong absorption maxima to 200 cm^{-1}; the single weak band near 250 cm^{-1} in Zn_2PtO_4 and Mg_2PtO_4 occurs below 200 cm^{-1} in most other spinels. White and De Angelis (20) have shown that, contrary to some predictions, inverse spinels should not and do not have more complex spectra than the normal spinels. Since the two types of ions in the octahedral sites are randomly distributed, the only result will be a shift in the peak frequencies rather than a removal of degeneracy. Table III lists tentative assignments given to the maxima in the zinc and magnesium spinels. They are in agreement with the conclusions reached by White and De Angelis in interpreting spinel spectra since internal stretching frequencies of MgO_4 tetrahedra in a number of spinels range from 690 to 565 cm^{-1}, and in ZnO_4 tetrahedra from 667 to 555 cm^{-1}.

Table III. Absorption Bands in Platinum Spinels

Mg_2PtO_4	Zn_2PtO_4	Assignment
655 cm^{-1}	612 cm^{-1}	Metal–oxygen stretch in tetrahedron
575	565	Lattice-stretching mode involving octahedron
490	465	Bending mode in tetrahedron
240	252	Lattice bending mode

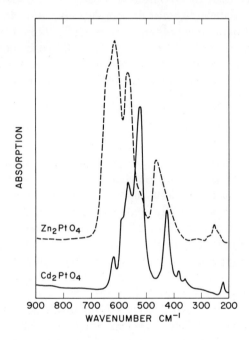

Figure 5. Infrared spectra of Zn_2PtO_4 and Cd_2PtO_4

The high frequency vibration is not assigned to a PtO_6 octahedral stretching mode because the mixed $(MPt)O_6$ grouping is expected to shift the vibration to a lower frequency.

Muller and Roy (8) have reported the synthesis of Cd_2PtO_4 and have assigned the Sr_2PbO_4 structure to this new compound. We have prepared this compound and obtained orange lath-like crystals at 1550°C and 40 kb pressure. Our powder data on Cd_2PtO_4 are in agreement with the Sr_2PbO_4 structure assignment; we hope to report a detailed structure analysis later. The infrared spectrum is presented in Figure 5. Peak assignments have not been attempted since insufficient structural data are available. Thermal analysis indicates decomposition of Cd_2PtO_4 to CdO and Pt at 840°C, followed by decomposition of CdO above 1000°C.

Our attempts to prepare comparable platinum spinels of the remaining divalent transition metal ions have not succeeded. We have indicated that MnO and FeO are oxidized by PtO_2, but no spinel formation with CoO, NiO, or CuO could be confirmed. Apparently these divalent ions cannot be forced into tetrahedral sites against their octahedral site preference energy. The mixed spinels $ZnCoPtO_4$, $ZnNiPtO_4$, and $MgCoPtO_4$ have, however, been prepared. In these compounds, cobalt and nickel can replace the octahedrally coordinated zinc and magnesium of the simple spinels.

MPt_3O_6 Compounds. Although CoO, NiO, and CuO do not react with PtO_2 to form spinels, they do participate in reactions with the dioxide to form ternary oxides having the general formula MPt_3O_6; ZnO, MgO, CdO, and HgO also form members of this series. It is apparent from the formula that not all of the platinum is present in the tetravalent state, and the reactions involved represent a departure from those encountered in the systems discussed previously. A typical equation may be written

$$2HgO + 5PtO_2 + Pt \rightarrow 2HgPt_3O_6 \tag{2}$$

In our initial experiments, HgO and PtO_2 were mixed in a 1:1 molar ratio and heated under 40 kb pressure to 800°C for 1 hour. A single phase crystalline product was recovered, purified in aqua regia, and found to be indexible as a hexagonal cell with $a = 7.25$ A and $c = 5.16$ A. The measured pycnometric density is 12.33 grams/cm^3. Thermal analysis indicated that the compound is not $HgPtO_3$ since decomposition at 650°C is accompanied by only a 33.5% weight loss. A complete analysis of the new compound gives the following percentages: oxygen 10.8%, mercury 22.6%, and platinum 66.6%. The calculated percentages for $HgPt_3O_6$ are: oxygen 10.9%, mercury 22.7%, and platinum 66.4%. In this instance, the Pt foil envelope is not inert to the reactants and participates in the reaction. Subsequent experiments with gold envelopes and with powdered platinum in the reaction mixture have confirmed the stoichiometry given in the equation above.

A similar compound is formed in the CdO–PtO_2 system, but the symmetry is pseudohexagonal orthorhombic. In addition, certain weak lines discernible in the powder pattern indicate that a larger cell must be chosen. A tentative indexing of $HgPt_3O_6$ and $CdPt_3O_6$ which gives reasonable agreement in intensities places these compounds in space group $Immm - D_{2h}^{25}$ with Hg or Cd in positions $2a$ and $2c$, divalent platinum in positions $2b$ and $2d$, tetravalent Pt in $8k$, O_I in $8m$, and O_{II} and O_{III} in $8n$. Although we have chosen arbitrary values for the oxygen coordinates based on a knowledge of cation radii, we have not attempted to verify the structure.

Table IV. Unit Cell Dimensions of MPt_3O_6 Compounds

Compound	Dimensions in A ($\pm$ 0.005)		
	a	b	c
$HgPt_3O_6$	10.334	7.264	6.286
$CdPt_3O_6$	10.189	7.215	6.327
$ZnPt_3O_6$	9.953	7.131	6.287
$MgPt_3O_6$	9.928	7.115	6.304
$NiPt_3O_6$	9.919	7.109	6.234
$CoPt_3O_6$	9.925	7.084	6.234

Table V. Powder Data for $MgPt_3O_6$ (Cr Radiation)

orthorhombic a = 9.928, b = 7.115, c = 6.304 A

hkl	d(A)	I_{obsd}	hkl	d	I
110	5.733	20	004	1.573	45
200	4.916	10	042	1.544	5
020	3.527	5	114	1.518	10
002	3.132	1	620	1.4969	70
310	2.984	15	242	1.4760	85
220	2.878	100	602	1.4626	65
112	2.754	15	440	1.4433	70
202	2.651	65	150	1.4065	5
400	2.470	50	314	1.3949	5
022	2.350	55	224	1.3817	80
130	2.298	1	532	1.3681	10
312	2.168	5	622	1.3528	10
222	2.123	10	404	1.3286	60
420	2.025	5	442	1.3127	5
402	1.945	5	152	1.2842	30
330	1.921	10	712	1.2718	30
510	1.907	5	424	1.2459	5
132	1.855	5	800	1.2402	80
040	1.774	60	334	1.2188	35
422	1.706	85	352	1.2061	20
240	1.669	5	044	1.1796	100
332	1.639	5	820	1.1708	20
512	1.631	5	260	1.1532	100

Analogous compounds are obtained with the transition metal monoxides. Here again, our early results suggested a lower platinum content for the orthorhombic phases, but a recognition of the participation of Pt metal in the reaction led to a revision of the composition. Subsequent experiments at the stoichiometry given by Equation 2, together with chemical analyses of the purified products, have confirmed the MPt_3O_6

formulation for the entire series. Cell dimensions for the six compounds are given in Table IV and powder data for the magnesium compound in Table V. The copper compound is not listed since it is not isostructural with the other members of the series, presumably owing to a Jahn-Teller distortion. Muller and Roy (7) have reported some structural data on $CuPt_3O_6$.

Thermal analysis of the MPt_3O_6 series indicates a somewhat diminished stability relative to the M_2PtO_4 compounds; decomposition to MO and Pt occurs at 650°–700°C, which is about 150° lower than the spinel decomposition temperature.

Acknowledgment

We are indebted to B. S. Tani for assistance in acquiring the x-ray data, and to Vaughn Leigh, Fleet Rust, and John Beres for their help with some of the experiments at high pressure.

Literature Cited

(1) Busch, R. H., Galloni, E. E., Raskovan, J., Cairo, A. E., *Anais Acad. Brasil. Cienc.* 1952, 24, 185.
(2) Hoekstra, H. R., *Inorg. Chem.* 1966, 5, 754.
(3) Hoekstra, H. R., Gallagher, F., *Inorg. Chem.* 1968, 7, 2553.
(4) Hoekstra, H. R., Siegel, S., *Inorg. Chem.* 1968, 7, 141.
(5) Muller, O., Roy, R., *Mater. Res. Bull.* 1969, 4, 39.
(6) Muller, O., Roy, R., *J. Less Common Metals* 1968, 16, 129.
(7) Muller, O., Roy, R., *J. Less Common Metals* 1969, 19, 209.
(8) Muller, O., Roy, R., *J. Less Common Metals* 1970, 20, 161.
(9) Pinsker, Z. G., *J. Phys. Chem. USSR* 1941, 15, 559.
(10) Randall, J. J., Ward, R., *J. Am. Chem. Soc.* 1959, 81, 2629.
(11) Randi, G., *Atti Accad. Naz. Lincei, Rend., Classe Sci. Fis. Mat. Nat.* 1966, 41, 197.
(12) Scheer, J. J., Van Arkel, A. E., Hayding, R. D., *Can. J. Chem.* 1955, 33, 683.
(13) Shannon, R. D., *Solid State Comm.* 1968, 6, 139.
(14) Shannon, R. D., Prewitt, C. T., *Acta Cryst.* 1969, B25, 925.
(15) Siegel, S., Hoekstra, H., Sherry, E., *Acta Cryst.* 1966, 20, 292.
(16) Siegel, S., Hoekstra, H. R., Tani, B. S., *J. Inorg. Nucl. Chem.* 1969, 31, 3803.
(17) Sleight, A. W., *Mater. Res. Bull.* 1968, 3, 699.
(18) Statton, W. O., *J. Chem. Phys.* 1951, 19, 40.
(19) Sundholm, A., Anderson, S., Magneli, A., Marinder, B., *Acta Chem. Scand.* 1958, 12, 1343; Marinder, B., Magneli, A., *ibid.*, 1958, 12, 1345.
(20) White, W. B., De Angelis, B. A., *Spectrochim. Acta* 1967, 23A, 985.

RECEIVED December 16, 1969. Work performed under the auspices of the U. S. Atomic Energy Commission.

5

Nitrido Complexes of the Platinum Group Metals

M. J. CLEARE and F. M. LEVER

Johnson, Matthey & Co., Ltd., Research Laboratories,
Wembley, Middlesex, England

W. P. GRIFFITH

Inorganic Research Laboratories, Imperial College of Science & Technology,
London, S.W. 7, England

> *A new class of binuclear nitrido complexes of tetravalent osmium and ruthenium is described in which the metal atoms are symmetrically bridged by a nitride ligand to give a linear M–N–M unit. They have the stoichiometries $[M_2NX_8(H_2O)_2]^{3-}$ and $[M_2N(NH_3)_8Y_2]^{3+}$ (M = Os, Ru; X = Cl, Br; Y = Cl, Br, etc.). Studies are reported on their vibrational spectra, structures, and bonding. Preliminary studies are reported also on trinuclear complexes of osmium and iridium. Finally, the use of vibrational spectroscopy in the study of metal–nitrido and metal–oxo complexes is discussed briefly.*

The nitride ion, N^{3-}, is a strong π-donor ligand and, as such, one expects to find it complexed with a metal in a relatively high oxidation state, similar to oxide and fluoride in their complexes. Of the six platinum group metals, nitrido complexes are known at present only for osmium, ruthenium, and iridium. These are shown in Table I. The terminal complexes of osmium have been known for many years, and we intend in this paper to discuss them only briefly, concentrating in more detail on the binuclear species of osmium and ruthenium, which we have been studying recently. The trinuclear complexes of iridium and osmium are discussed in the light of recent spectroscopic data. A heterometallic species, $(PEt_2Ph)_3Cl_2ReNPtCl_2(PEt_3)$, with a nitrogen bridge between a platinum and a rhenium atom has been described recently by Chatt and Heaton (2), but will not be discussed here.

Terminal Osmium Nitrido Complexes

These compounds have been known for many years and the preparative methods are indicated in Table II.

Potassium osmiamate, $K[Os(VIII)O_3N]$, is prepared by the addition of aqueous ammonia to a solution of osmium tetroxide, OsO_4, in caustic potash solution (6). X-ray studies show the molecule to be basically

Table I. Nitrido Complexes of the Platinum Metals

Osmium

Terminal	Os(VIII) $[OsO_3N]^-$
	Os(VI) $[OsNX_5]^{2-}$ $_5[OsNX_4H_2O]^-$
	(X = Cl,Br,CN,1/2 OX)
Bridging	Os(IV) $[Os_2NX_8(H_2O)_2]^{3-}$ (X = Cl,Br)
	$[Os_2N(NH_3)_8X_2]^{3+}$
	(X = Cl,Br,I,NCS,N_3,NO_3)
Trinuclear	$Os_3N_7O_9H_{21}$

Ruthenium

Bridging only	$[Ru(IV)_2N\ X_8(H_2O)_2]^{3-}$ (X = Cl,Br,NCS)
	$[Ru(IV)_2N(NO_2)_6(OH)_2(H_2O)_2]^{3-}$
	$[Ru(IV)_2N(NH_3)_8\ X_2]^{3+}$ (X = Cl,Br,NO_3)
	$[Ru(IV)_2N(NH_3)_6\ (H_2O)\ X_3]^{2+}$ (X = Cl,NCS,N_3)

Iridium

Trinuclear only:	$[Ir_3N(SO_4)_6(H_2O)_3]^{4-}$
	$[Ir_3N(SO_4)_6(OH)_3]^{7-}$
	$[Ir_3N\ Cl_{12}(H_2O)_3]^{4-}$

Table II. Terminal Osmium Nitrido Complexes

Os(VIII)

$$OsO_4 + OH^\ominus + NH_3 \rightarrow [OsO_3N]^- + 2H_2O$$

Potassium osmiamate forms pale yellow crystals.
Molecular structure: distorted tetrahedral

$$\nu\ Os \equiv N\ 1021\ cm^{-1}.$$

Os(VI)

$$[OsO_3N]^- + X^- + 6HX \rightarrow [Os\ N\ X_5]^{2-} + X_2 + 3H_2O$$

$$(X = Cl,Br)$$

These complexes tend to lose the ligand trans to the nitride to give (Os $NX_4H_2O)^-$ (X = Br, CN, 1/2 ox) in aqueous solution; $\nu\ Os \equiv N \sim 1080\ cm^{-1}$.

Molecular structure: distorted octahedral (C_{4v}) with the osmium atom slightly out of the X_4 plane towards the nitride (in the case of $[OsNCl_5]^{3-}$).

tetrahedral, although somewhat distorted (12). The Os≡N stretching frequency is at 1021 cm^{-1}, which shifts to 993 on ^{15}N substitution (16).

Os(VI) complexes of type [Os(VI)NX$_5$]$^{2-}$ or [Os(VI)NX$_4$H$_2$O]$^-$ (X = Cl, Br, CN, ½ox) are made by the action of the appropriate acid HX on potassium osmiamate (6, 11). The tendency to lose the ligand trans to the nitride and to substitute a water molecule is some evidence of a trans labilizing effect of π-donor ligands. X-ray studies show a distorted octahedral structure with the osmium atom slightly out of the plane of the equatorial X atoms towards the nitride (1). The Os≡N stretching frequency is around 1080 cm^{-1} (11) in such complexes.

Binuclear Nitrides of Osmium and Ruthenium

In general, there is a similar range of complexes for each metal, both being in the oxidation state (IV).

Anionic Complexes. The preparation of these compounds is indicated in Table III. The product of the reaction of (NH$_4$)$_2$[OsCl$_6$] with chlorine gas at 400°C will dissolve in dilute hydrochloric acid; from this

Table III. Binuclear Osmium and Ruthenium Nitrido Complexes

Anionic Complexes

$$(NH_4)_2[OsCl_6] \xrightarrow[\text{then HCl}]{Cl_2, 400°C} [Os_2NCl_8(H_2O)_2]^{3-}$$

$$[RuNOCl_5]^{2-} \begin{array}{c} \xrightarrow{SnCl/HCl} [Ru_2NCl_8(H_2O)_2]^{3-} \\ \xrightarrow{HCHO/OH^-} [RuN(OH)_5 \cdot nH_2O] \end{array}$$

$$[RuO_4]^{2-} \xrightarrow{NH_3 \text{ Aq.}/OH^\ominus} [RuN(OH)_5 \cdot nH_2O]$$

$$[Ru_2N(OH)_5 \cdot nH_2O] \xrightarrow{HBr} [Ru_2NBr_8(H_2O)_2]^{3-}$$

$$[Ru_2N(OH)_5 \cdot nH_2O] \xrightarrow{HCl} [Ru_2NCl_8(H_2O)_2]^{3-}$$

$$[Ru_2N(NCS)_8(H_2O)_2]^{3-} \xleftarrow{NCS^\ominus} \quad \downarrow NO_2^\ominus$$

$$[Ru_2N(NO_2)_6(OH)_2(H_2O)_2]^{3-}$$

solution, salts of the type $M(I)_3[Os(IV)_2NCl_8(H_2O)_2]$ can be precipitated (5). The analogous ruthenium complexes may be prepared by reaction of $K_2[Ru(NO)Cl_5]$ with stannous chloride in hydrochloric acid solution (4). Other products are formed in this reaction, in particular a nitrosyl complex containing coordinated $SnCl_3^-$, which is still under investigation.

Addition of alkali to an aqueous solution of $[Ru(IV)_2NCl_8(H_2O)_2]^{3-}$ yields a brown gelatinous precipitate of $Ru_2N(OH)_5 \cdot nH_2O$ which will redissolve in hydrochloric acid to regenerate the original complex. This hydroxy species may be prepared directly by the action of formaldehyde with $K_2[Ru(NO)Cl_5]$ in alkaline solution, or by reaction of potassium ruthenate, $K_2[RuO_4]$ with excess aqueous ammonia. Hydrobromic acid on the hydroxy complex yields $[Ru_2NBr_8(H_2O)_2]^{3-}$, while the action of excess thiocyanate on a hydrochloric acid solution of the hydroxide gives $[Ru_2N(NCS)_8(H_2O)_2]^{3-}$. Potassium nitrite reacts with an aqueous solution of $[Ru_2NCl_8(H_2O)_2]^{3-}$ to give $[Ru_2N(NO_2)_6(OH)_2(H_2O)_2]^{3-}$. Infrared spectra indicate that both of the latter ligands (NO_2 and NCS) are N-bonded to the metal (5).

Molar conductances of these anionic complexes in aqueous solution are initially close to the values expected for 3:1 electrolytes, but on long standing or on heating the conductances increase greatly, and there is a parallel decrease in the pH of the solution. We assume that this results from replacement of the halo ligands by aquo groups and loss of a proton from the water ligands. The effect is less marked for the osmium complexes than for ruthenium, presumably because of the greater inertness of third-row elements towards substitution (5).

Cationic Complexes. One series of ammine complexes has been found for osmium ($[Os_2N(NH_3)_8Y_2]^{3+}$) and two for ruthenium ($[Ru_2N(NH_3)_8X_2]^{3+}$ and $[Ru_2N(NH_3)_6(H_2O)X_3]^{2+}$). The preparative methods are indicated in Table IV.

Reaction of sodium chloroosmate $Na_2[OsCl_6]$ with aqueous ammonia at $\sim 100°C$ in a Carius tube yields $[Os_2N(NH_3)_8Cl_2]Cl_3$. Another method is the reaction of ammonia with $K_3[Os_2NCl_8(H_2O)_2]$. The product will react with excess X^- ion to give $[Os_2N(NH_3)_8X_2]X_3$ ($X = Br^-, I^-, NCS^-, N_3^-, NO_3^-$) after short refluxing in aqueous solution. Salts of the form $[Os_2N(NH_3)_8X_2]Y_3$ may be prepared metathetically with Y^- ion in the cold, suggesting that only two of the five X groups are coordinated.

The reaction of aqueous ammonia on $K_3[Ru_2NX_8(H_2O)_2]$ gives a mixture of $[Ru_2N(NH_3)_8X_2]X_3$ and $[Ru_2N(NH_3)_6(H_2O)X_3]X_2$ ($X = Cl, Br$). The former yields species of the form $[Ru_2N(NH_3)_8X_2]Y_3$ on metathesis in the cold with Y^- ion ($Y = Br, I, NCS, N_3, NO_3^-$), while boiling with an excess of certain ligands results in ammine replacement and conversion to a hexammine species—*i.e.*, $[Ru_2N(NH_3)_6(H_2O)Y_3]Y_2$ (Y

Table IV. Binuclear Osmium and Ruthenium Nitrido Complexes

Cationic Complexes

$$[OsX_6]^{2-} \xrightarrow[\text{100°C, Carius tube}]{\text{NH}_3 \text{ Aq.}} [Os_2N(NH_3)_8 X_2] X_3 \quad (X = Cl, Br)$$

$$[Os_2NCl_8(H_2O)_2]^{3-} \xrightarrow[\text{100°C, Carius tube}]{\text{NH}_3 \text{ Aq.}} [Os_2N(NH_3)_8 Cl_2]Cl_3$$

$$[Os_2N(NH_3)_8 Cl_2]^{3+} \xrightarrow[\text{Reflux in aq. soln}]{\text{Excess X}^-} [Os_2N(NH_3)_8 X_2]^{3+}$$

$$(X = Br, I, NCS, N_3, NO_3)$$

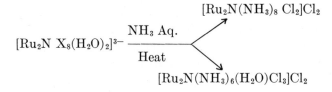

$$[Ru_2N X_8(H_2O)_2]^{3-} \xrightarrow[\text{Heat}]{\text{NH}_3 \text{ Aq.}} \begin{matrix} [Ru_2N(NH_3)_8 Cl_2]Cl_2 \\ [Ru_2N(NH_3)_6(H_2O)Cl_3]Cl_2 \end{matrix}$$

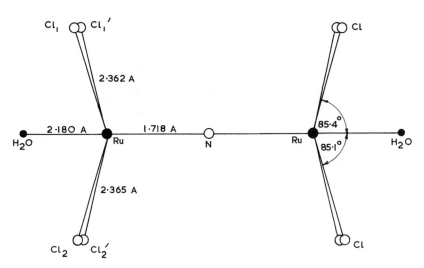

Figure 1. Structure of the anion in $K_3[Ru_2NCl_8(H_2O)_2]$

$= N_3^-, NCS^-, NO_2^-$). It appears that other ligands—e.g., NO_3^-—will leave the compound as an octammine. Species of the type $[Ru_2N(NH_3)_6(H_2O)Y_3]Z_2$ may be made by metathesis with Z^- in the cold ($Z = Cl, I$).

Molar conductances for the osmium and ruthenium octammines in aqueous solution were close to the values expected for 3:1 electrolytes (~360 mhos cm^2) but rose on standing or heating to give final values nearer that expected for a 5:1 electrolyte (~600 mhos cm^2). This is presumably because of trans aquation; the effect was again less marked for osmium. Significantly, all the halogen in the halo species [Ru$_2$N(NH$_3$)$_8$-X$_2$]X$_3$ could be precipitated with silver nitrate from heated solutions. Conductivities for the hexammines also varied somewhat with time, but initial values were lower than for the octammines and indicated the presence of a 2:1 electrolyte (240 mhos cm^2) (5)

Structure. An x-ray structural determination has been carried out on K$_3$[Ru$_2$NCl$_8$(H$_2$O)$_2$] (3) and the results are depicted in Figure 1. The anion has a linear skeleton with the eight equatorial chlorine atoms eclipsed so that the over-all idealized symmetry of the anion is D_{4h}. Each ruthenium atom is displaced 0.19 A out of the Cl$_4$ plane toward the nitrogen atom. Such a displacement of metal atoms towards a π-donor ligand has been observed in other complexes, such as K$_4$[Re$_2$OCl$_{10}$] (18) and (Ph$_4$As)[MoOBr$_4$H$_2$O] (7), as well as K$_2$[OsNCl$_5$] (1), which has been mentioned previously. It has been suggested that this arises mainly from the repulsion between the nitrogen and chlorine ligands within the molecule.

Vibrational Spectra. Spectra of typical complexes are listed in Tables V and VI. The compounds exhibit similar spectra, suggesting that they all have the same basic structure. The main features are:

(a) The presence of a strong sharp band in the infrared in the range 1050–1130 cm^{-1}, not observed in the Raman. This band is little changed in frequency on deuteriation, but shifts downward by some 30 cm^{-1} on

Table V. Vibrational Spectra of Bridging Nitrido Complexes—Anionic Species

$(M_2NX_8(H_2O)_2)^{3-}$		$\nu_{M_2N}{}^{as}$	$\nu_{2MN}{}^{s}$	ν_{M-X}	δ_{X-M-X}
K$_3$(Os$_2$N Cl$_8$(H$_2$O)$_2$)	R	—	267(10)	{315(4) {295(2)	185w
	IR	1137 vs	—	303s br	
	^{a}IR	1135 vs	—		
K$_3$(Ru$_2$N Cl$_8$(H$_2$O)$_2$)	R	—	329(4)	{301(10) {294(4)	200w
	^{a}R	—	330(4)p		
	IR	1078 vs	—	{315s {289m	
	^{a}IR	1080 vs	—		
K$_3$(Ru$_2{}^{15}$N Cl$_8$(H$_2$O)$_2$)	IR	1046 vs	—	{315s {289m	

^{a} Indicates aqueous solution measurement.

Table VI. Vibrational Spectra of Bridging Nitrido Complexes—Cationic Species

$(M_2N(NH_3)_8X_2)^{3+}$	$\nu_{M_2N}{}^{as}$	$\nu_{M_2N}{}^{s}$	ν_{M-NH}	δ_{M-NH}	ν_{M-X}
$(Os_2N(NH_3)_8Cl_2)Cl_3$	R — IR 1104 vs	299(10) —	455w 465w	267(1) 260s br	286(3) 282m
$(Os_2N(ND_3)_8Cl_2)Cl_3$	R — IR 1101 vs	286(10) —	430w 440w	254(1) 240s br	278m 281m
$(Ru_2N(NH_3)_8Cl_2)Cl_3$	R — IR 1055	354(1) —	{496(3) {475(1) 465w	265w br 258m br	280w 280m sh
$(Ru_2N(ND_3)_8Cl_2)Cl_3$	R — IR 1054 vs	339(10) —	440vw 430w	250(1) 245s br	273w 277m sh

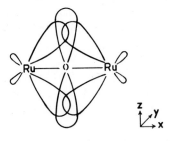

Figure 2. Bonding in Ru—O—Ru and Ru—N—Ru systems

Diagram shows $4d_{xz}$–$2p_z$–$4d_{xz}$ overlap. There will be a similar $4d_{xy}$–$2p_y$–$4d_{xy}$ overlap at right angles to the plane of the paper.

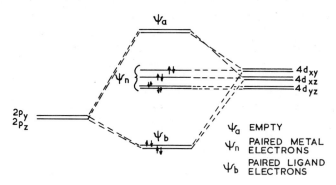

Figure 3. Bonding in bridging nitride complexes

The M.O. diagram indicates the formation of two sets of 3-center molecular orbitals with the metal d electrons paired in the nonbonding orbitals

^{15}N substitution in $[Os_2{}^{15}N^{15}(NH_3)_8Br_2]Br_3$ and $K_3[Ru_2{}^{15}NCl_8(H_2O)_2]$. This band we have assigned to $\nu^{as}{}_{M_2N}$, the asymmetric M—N—M stretch. This mode consists almost entirely of nitrogen movement and is expected to occur in the same region as the terminal metal nitride stretches.

(b) The presence of a strong, sharp, polarized Raman band near 350 cm^{-1} for the ruthenium complexes and 280 cm^{-1} for the osmium complexes, which we assign to the symmetric M—N—M stretch, $\nu^s_{M_2N}$. These bands are not observed in the infrared spectra. They shift downward some 13 cm^{-1} on deuteriation of the ammine complexes, but this is likely to be caused by strong coupling of $\nu^s_{M_2N}$ (a totally symmetric mode) with the symmetric metal ammine stretches and deformations having the same symmetry species (A_{1g}), together with a slight effect arising from the greater mass of deuterium. Here the mode consists mainly of metal movement; this is clearly shown by the effect of the extra mass of osmium. The general lack of coincidence between Raman and infrared bands and, in particular, the fact that the Raman-active $\nu^s_{M_2N}$ and infrared-active $\nu^{as}_{M_2N}$ are inactive in the infrared and Raman, respectively, strongly suggest the presence of a center of symmetry. Thus, as in [Ru$_2$NCl$_8$(H$_2$O)$_2$]$^{3-}$ the aquo groups lie trans to the nitride group, it is likely that the X groups in [M$_2$N(NH$_3$)$_8$X$_2$]$^{3+}$ will be placed similarly (5).

Electronic Spectra. All these complexes are diamagnetic as a result of the π-bonding, which is shown in Figure 2.

Of the four metal d electrons on each ruthenium atom (Ru(IV)d^4), two (d_{xy},d_{xz}) will pair up in the nonbonding molecular orbital of the set produced by the 3-center overlap between the d_{xy} and d_{xz} orbitals on each metal atom, and the $2p_y$ and $2p_z$ orbitals on the nitride. The lower energy nitride electrons will pair in the bonding molecular orbitals (Figure 3). The remaining two electrons per ruthenium atom are paired in the 4 d_{yz} orbital which remains essentially nonbonding. It is a consequence of this bonding scheme that the M—N—M group should be linear and that the equatorial ligands should be in the eclipsed (D_{4h}) rather than the staggered (D_{4d}) positions.

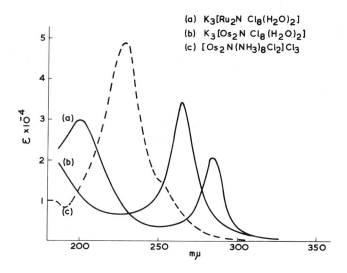

Figure 4. Electronic spectra of some bridging nitrido complexes

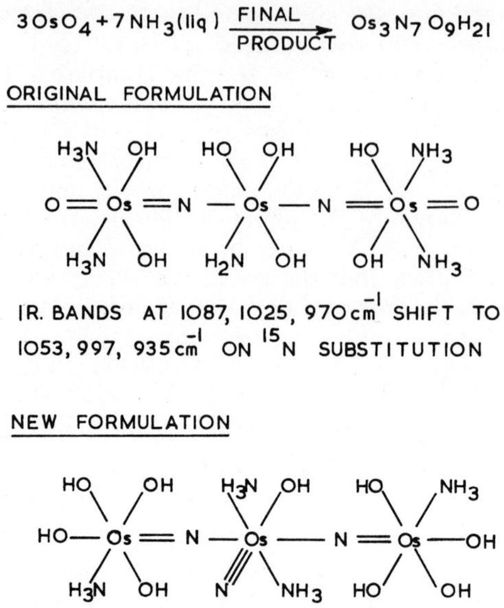

$3 OsO_4 + 7 NH_3 (liq) \xrightarrow[\text{PRODUCT}]{\text{FINAL}} Os_3N_7O_9H_{21}$

ORIGINAL FORMULATION

```
    H3N  OH       HO  OH        HO   NH3
      \ /           \ /           \ /
  O = Os = N  ― Os ― N  = Os = O
      / \           / \           / \
    H3N  OH      H2N  OH        OH   NH3
```

IR. BANDS AT 1087, 1025, 970 cm^{-1} SHIFT TO
1053, 997, 935 cm^{-1} ON ^{15}N SUBSTITUTION

NEW FORMULATION

```
    HO   OH       H3N  OH        HO   NH3
      \ /           \ /           \ /
  HO― Os = N ― Os ――― N = Os ― OH
      / \           ⫽ \           / \
    H3N  OH        N  NH3       HO   OH
```

Figure 5. Trinuclear Os(VI) nitrido complex

The electronic spectra (350–190 mμ) of three of these complexes are shown in Figure 4. Jorgensen and Orgel (*14*) have suggested that similar bands observed in $K_4[Ru_2OCl_{10}]$ arise from transitions from the halogen ligands to the antibonding molecular orbital from the 3-center bond. We follow this assignment for the $[M_2NX_8Y_2]$ system, although it is not strictly isoelectronic with $[Ru_2OX_{10}]^{4-}$. The Ru—O—Ru and Ru—N—Ru groupings are, however, isoelectronic. The energy of the antibonding molecular orbital is determined largely by the extent of the Ru—O—Ru or Ru—N—Ru π-bonding. The bands are seen near 400 mμ for $[Ru_2OCl_{10}]^{4-}$ and at 397 and 330 mμ in $(NH_4)_2[Os_2OCl_{10}]$. In nitride they are higher in energy (*e.g.*, $K_3[Os_2NCl_8(H_2O)_2]$, 273 mμ; $K_3[Ru_2NCl_8(H_2O)_2]$, 294 mμ), indicating that M–N π-bonding is stronger than for M–O in these binuclear systems.

Trinuclear Osmium Complex

The final product of the reaction of liquid ammonia with OsO_4 has the empirical formula $Os_3N_7O_9H_{21}$ (*17, 20*). We have studied the infrared spectrum of the ^{15}N-substituted compound and propose structure II (Figure 5) rather than structure I, which was suggested originally. Bands at 1087, 1025, and 970 cm^{-1} shift to 1053, 997, and 935 on ^{15}N-substitution and may be assigned to Os—N—Os or Os—N stretching modes. The new formulation accounts for the observation that only two molecules of ammonia are released on heating with alkali, since two of the ammine groups are in a different environment from the other two (5), and thus presumably have different stabilities in respect to metal–nitrogen bond strengths.

Trinuclear Iridium Complexes

The reaction of $(NH_4)_3[IrCl_6]$ with boiling sulfuric acid yields a solution from which green salts of the form $M_4[Ir_3N(SO_4)_6(H_2O)_3]$ can be precipitated (8, 15). It has been suggested that the anion has the structure shown on Figure 6 with a coplanar Ir_3N unit and π-bonding between the 2 p_z orbital (perpendicular to the Ir_3N triangle) and the iridium atoms (13, 19). The oxidizing power of the species is consistent with it's containing one Ir(III) and two Ir(IV) atoms per molecule (8, 13). Reduction with vanadous ion gives a straw-yellow species containing three Ir(III) atoms per molecule. X-ray studies on the somewhat similar $[Cr_3O(CH_3COO)_6(H_2O)_3]Cl \cdot 5H_2O$ have shown that the Cr_3O unit is planar with a triangular arrangement of metal atoms (9).

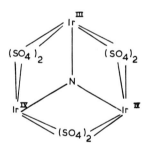

Figure 6. Proposed structure for iridium trinuclear nitrides

The Ir_3N unit is believed to be coplanar with the iridium atoms at the corners of an equilateral triangle. A related structure has been found for $[Cr_3O(Acetate)_6]OAc$.

Table VII. Trinuclear Iridium Nitrido Complexes

$$3(NH_4)_3[IrCl_6] \xrightarrow[H_2O]{H_2SO_4} (NH_4)_4[Ir_3N(SO_4)_6(H_2O)_3]$$

$$[Ir_3N(SO_4)_6(H_2O)_3]^{4-} \xrightarrow[\text{Cold}]{KOH} K_7[Ir_3N(SO_4)_6(OH)_3]$$

$$[Ir_3N(SO_4)_6(H_2O)_3]^{4-} \xrightarrow[\text{Boiling}]{\text{Caustic alkali}} Ir_3N(OH)_{11} \cdot nH_2O \xrightarrow{\substack{HCl \\ +CsCl}} Cs_4[Ir_3NCl_{12}(H_2O)_3]$$

Other reactions of $[Ir_3N(SO_4)_6(H_2O)_3]^{4-}$ are shown on Table VII. Cold caustic alkalies react to replace the aquo groups, giving $[Ir_3N(SO_4)_6(OH)_3]^{7-}$, while hot alkali appears to replace the sulfato ligands to give a hydroxy precipitate which probably may be described as $Ir_3N(OH)_{11} \cdot nH_2O$. The Ir_3N group is still intact as reaction with sulfuric acid regenerates the original complex. The hydroxy complex dissolves in HCl to give a species which appears to be $[Ir_3NCl_{12}(H_2O)_3]^{4-}$.

We have measured the infrared spectra of these species containing both ^{14}N and ^{15}N, and in each case a band near 780 cm^{-1} shifts downward in frequency by some 20 cm^{-1} on ^{15}N substitution. We assign this band to the asymmetric Ir_3N stretch. The Raman spectra of the complexes were poor, owing to their unfavorable colors, but weak bands near 230 cm^{-1} may be caused by the symmetric Ir_3N stretch (5). The complex splitting of the sulfate modes is similar to that observed in $K_{10}[Ir_3O(SO_4)_9]$ (10), and presumably arises from the low site symmetry of the SO_4 groups which function here as bidentate ligands.

Table VIII. Vibrational Spectra of Oxy and Nitrido Complexes

	$X = O$, cm^{-1}	$X = N$, cm^{-1}
TERMINAL M–X		
ν_{MX}	900–1060	950–1180
δ_{MX}	300–400	ca. 350
BRIDGING M_mX_n		
Linear M–X–M		
$\nu_{M_2X}{}^{as}$	760–880	1000–1160
$\nu_{M_2X}{}^{s}$	200–270	260–370
Linear M–X–M–X–M		
$\nu_{M_3X_2}{}^{as}$	ca. 820	ca. 1000
$\nu_{M_3X_2}{}^{s}$	ca. 220	–
Triangular M_3X		
$\nu_{M_3X}{}^{as}$	500–650	700–800
$\nu_{M_3X}{}^{s}$	180–240	ca. 230

Comparison of the Vibrational Spectra of Oxy and Nitrido Complexes

Table VIII shows the characteristic ranges of frequencies for stretching and deformation modes for terminal and bridging oxy and nitrido systems. In each case, the frequencies for the stretching modes are higher for nitrides than for oxides. Although the nitrogen atom is lighter than oxygen, it is likely that this effect is caused in large part by the fact that the nitride ligand is a more efficient π-donor than oxide—it is less electronegative and has more negative charge to impart to the metal atom (5).

Literature Cited

(1) Bright, D., Ibers, J. A., *Inorg. Chem.* **1969**, 8, 709.
(2) Chatt, J., Heaton, B. T., *Chem. Commun.* **1968**, 274.
(3) Ciechanowicz, M., Skapski, A. C., *Chem. Commun.* **1969**, 574.
(4) Cleare, M. J., Griffith, W. P., *Chem. Commun.* **1968**, 1302.
(5) Cleare, M. J., Griffith, W. P., *J. Chem. Soc.* **1970**, A, 1117.
(6) Clifford, A. F., Kobayashi, C. S., *Inorg. Syn.* **1960**, 6, 204.
(7) Cotton, F. A., Lippard, S. J., *Inorg. Chem.* **1965**, 4, 1621.
(8) Delepine, M., *Ann. Chim. (Paris)* **1959**, 1115, 1131.
(9) Figgis, B. N., Robertson, E. B., *Nature* **1965**, 205, 695.
(10) Griffith, W. P., *J. Chem. Soc.* **1969**, A, 2270.
(11) Griffith, W. P., *J. Chem. Soc.* **1965**, 3694.
(12) Jaeger, F. M., Zanstra, J. E., *Proc. Acad. Sci. Amsterdam* **1932**, 35, 610.
(13) Jorgensen, C. K., *Acta Chem. Scand.* **1959**, 13, 196.
(14) Jorgensen, C. K., Orgel, L. E., *Mol. Phys.* **1961**, 4, 215.
(15) Lecoq de Boisbaudran, *Compt. Rend.* **1883**, 96, 1336, 1406, 1551.
(16) Lewis, J., Wilkinson, G., *J. Inorg. Nucl. Chem.* **1958**, 6, 12.
(17) McCordie, W. C., Watt, G. W., *J. Inorg. Nucl. Chem.* **1965**, 27, 1130, 2013.
(18) Morrow, J. C., *Acta Cryst.* **1962**, 15, 851.
(19) Orgel, L. E., *Nature* **1960**, 187, 505.
(20) Potrafke, E. M., Watt, G. W., *J. Inorg. Nucl. Chem.* **1961**, 17, 248.

RECEIVED December 16, 1969.

6

Hydrido Complexes of Platinum Group Metals

L. M. VENANZI

State University of New York at Albany, Albany, N. Y. 12203

The preparation and properties of hydrido complexes of platinum group metals are reviewed briefly. The reactions of the more common hydrido complexes of iridium are schematically shown. The reactions of complexes $[IrHX_2$-$(Ph_3P)_3]$ ($X = Cl$, Br, and I), I, with the quadridentate ligand $(o\text{-}Ph_2P \cdot C_6H_4)_3P$, QP, are described and rationalized on the basis of the following reaction sequence:

$$\text{I} \xrightarrow{-HX} [IrX(Ph_3P)_3] \xrightarrow{\text{heat}} [IrHX(o\text{-}C_6H_4PPh_2)(Ph_3P)_2]$$
$$\xrightarrow{QP} [IrH(o\text{-}C_6H_4PPh_2)(QP)]X \xrightarrow{+HX} [IrHX(Ph_3P)(QP)]X$$

Complexes of the latter two types have been assigned seven-coordinate structures on the basis of their proton and phosphorus-31 NMR spectra even though the oxidation number of the iridium atom is 3+.

All platinum group metals give hydrido complexes of general formula $M_nH_mX_pL_q$ (X = anionic ligand and L = uncharged ligand). Although this field of chemistry is relatively new, its main development occurring over the last 10 years, the number of papers describing new compounds, their properties, and their reactions is very large and is growing at an increasing rate. Hydrido complexes also have been the subject of several reviews (9, 11, 23, 24, 25, 39), and the present one, like its predecessors, will be sadly out of date by the time it appears in print.

As its title implies, this review restricts itself to describing and discussing compounds of platinum group metals—i.e., of ruthenium, rhodium, palladium, osmium, iridium, and platinum—although the compounds of the other transition elements and even some post-transition elements are either fully analogous or closely related to those of the platinum metals.

Table I. Some Better Known Types of Hydrido Complex of the Platinum Metals[a]

Square Planar

trans-[MHXL$_2$]   [PtH$_2$L$_2$]
[IrH$_2$(LL)]   [Ir-HX(LL)]
[M'(LL)$_2$]X

Five-Coordinate

[IrH$_2$L$_3$]X   [IrH$_3$L$_2$]
[M'H(CO)L$_3$]   [IrH(CO)$_2$L$_2$]

Octahedral

[PtH$_2$X$_2$L$_2$]   [IrH$_2$L$_2$'L$_2$]X
[M'HX$_2$(CO)L$_2$]   [IrH$_2$L'L$_3$]X
[M"H$_n$X$_{2-n}$(LL)$_2$]   [M'H$_m$X$_{3-m}$L$_3$]
[M"HX(CO)L$_3$]

[a] M = Pd or Pt   M' = Rh or Ir
M" = Ru or Os   L = monodentate phosphine
L' = CO or monodentate phosphine   LL = bidentate phosphine
m = 1, 2, or 3   n = 1 or 2

Even within this limitation, the subject is vast and, therefore, no attempt is made to give a comprehensive description of the field.

A survey of the compounds reported in the literature shows that the formation of stable hydrido complexes generally is linked with the presence in the molecule of one or more of certain types of ligands, the most frequently occurring being carbon monoxide, phosphorus(III) and arsenic(III) derivatives, and the cyclopentadiene group.

It is useful to classify the complexes by their coordination number. The majority of them are six-coordinate with octahedral structure, although four- and five-coordinate species are numerous, and even some seven-coordinate species are known. Table I shows a list of the main types of compounds reported in the literature.

Preparative methods for hydrido complexes vary widely. Some of the more frequently used of them are:

(1) Action of complex hydrides, *e.g.*,

$$cis\text{-}[RuCl_2(Me_2PCH_2CH_2PMe_2)_2] \xrightarrow{LiAlH_4} trans\text{-}[RuHCl(Me_2PCH_2CH_2PMe_2)_2] \quad (16)$$

(2) Action of alcohols in the presence of a base, *e.g.*,

$$(NH_4)_2[OsBr_6] \xrightarrow[MeOCH_2CH_2OH]{Ph_3P} [OsHBr(CO)(Ph_3P)_3] \quad (41)$$

(3) Protonation of complexes, *e.g.*,

$$[Ru_2(\pi\text{-}C_5H_5)_2(CO)_6] \xrightarrow{H^+} [Ru_2H(\pi\text{-}C_5H_5)_2(CO)_6]^+ \quad (20)$$

(4) Action of hydrazine, e.g.,

$$cis\text{-}[PrCl_2(Et_3P)_2] \xrightarrow{N_2H_4 \cdot H_2O} trans\text{-}[PtHCl(Et_3P)_2] \quad (17)$$

Hydrido complexes are also formed in some unusual reactions:

(1) $PdCl_2 + CO + [Ph_4As]Cl \xrightarrow[\text{at room temp.}]{MeOCH_2CH_2OH} [Ph_4As][PdHCl_2(CO)] \quad (27)$

(2) $[PtI\,(MgI)(Pr_3P)_2] \xrightarrow{H_2O} [PtHI\,(Pr_3P)_2] \quad (19)$

Some of the preparative methods are best described, at least formally, as involving an "oxidative addition" of a molecule to a complex of a metal ion or atom in a low oxidation state. Examples of this type of reaction are shown below.

(1) Addition of molecular hydrogen, e.g.,

$$[Rh(Me_2PCH_2CH_2PMe_2)_2]Cl \xrightarrow{H_2} cis\text{-}[RhH_2(Me_2PCH_2CH_2PMe_2)_2]Cl \quad (12)$$

(2) Addition of hydrogen halide, e.g.,

$$[Rh(Ph_2PCH_2CH_2PPh_2)_2]Cl \xrightarrow{HCl} cis\text{-}[RhHCl(Ph_2PCH_2CH_2PPh_2)_2]Cl \quad (38)$$

(3) Addition of other hydrides, e.g.,

$$[Pt(Ph_3P)_3] \xrightarrow{H_2S} trans\text{-}[PtH(SH)(Ph_3P)_2] \quad (35)$$

A particularly interesting type of oxidative addition reaction is the "hydrogen abstraction" reaction, some examples of which are given below.

(1) $[Ru(Me_2PCH_2CH_2PMe_2)_2] \rightleftarrows [RuH(CH_2P(Me)CH_2CH_2PMe_2)$

$(Me_2PCH_2CH_2PMe_2)] \quad (14)$

(2) $[IrCl(Ph_3P)_3] \xrightarrow{heat} [IrHCl(o\text{-}C_6H_4PPh_2)(Ph_3P)_2] \quad (7)$

(3) $trans\text{-}[PtCl_2(Et_3P)_2] + LiCB_{10}H_{10}CMe \rightarrow [Pt(CB_{10}H_{10}CMe)$

$(CH_2CH_2PEt_2)(Et_3P)] \quad (8)$

These reactions are of particular interest as they could provide the basis for a number of catalytic processes involving saturated hydrocarbons.

The properties of hydrido complexes vary widely, ranging from those which can be detected only spectroscopically to those that show remarkable general chemical stability. Stability generally tends to be at a maximum in compounds of the third transition series. Furthermore, for a given metal and type of complex, stability appears to be at its highest when phosphine ligands are present in the molecule. Naturally, it would be foolhardy to take the above kind of stability as a guide to thermodynamic stability of the metal–hydrogen bond, as one would draw a correlation between a stability which is a composite of a number of thermodynamic and kinetic factors and a relatively small energy term such as the M–H bond energy. It is regrettable that thermodynamic data on hydrido complexes are totally lacking but they are almost impossible to obtain.

The most commonly quoted physical data on hydrido complexes are the vibrational spectra and proton magnetic resonance parameters. Hydrido complexes exhibit M–H stretching vibrations in the region 1700–2250 cm^{-1} and M–H bending modes in the region 660–850 cm^{-1}. The absence of absorption in the above regions, however, is not to be taken as an indication that the molecule under examination does not contain M–H bonds. Several hydrido complexes are known where M–H bonds were not detectable by infrared spectroscopy (28, 34, 37). Proton magnetic resonance, on the other hand, is a more generally useful technique. Proton resonances in hydrido complexes have very characteristic τ-values ranging from 11 to 42 ppm relative to TMS. The limitation here is introduced by solubility and by fast exchange phenomena although, in many cases, the former difficulty can be overcome by spectrum accumulation and the latter by lowering the sample temperature.

Hydrido ligands exhibit a very strong trans effect which appears to operate mainly by a labilization of the bond in trans position to the M–H bond. This labilization generally is associated with a lengthening of the bond which has been labilized, an effect operating mainly by trans influence (36) caused by a selective rehybridization of the metal σ orbitals leading to the M–H bond having a large d and s component. Such effects are most evident in square planar complexes, but they also have been observed in octahedral complexes.

The general reactivity of hydrido complexes is high, and it is related mainly to the coordination number of the central metal atom. Thus, four- and five-coordinate complexes usually are more reactive than related six-coordinate species. The latter, however, ordinarily are more reactive than the corresponding complexes which do not contain M–H bonds. Such reactivity may be induced either by operation of the trans effect or, where possible, by the reductive elimination of a molecule of hydrogen halide with or without assistance of a base.

The chemical behavior of M–H bonds varies from that of an active hydride to that of a strong acid. This wide variation in properties, however, may be achieved with a relatively small variation of charge distribution (4). Attempts to correlate chemical shifts and coupling constants with bond distances and electron distribution in M–H bonds have led to the conclusion that such correlations may be valid only when very small changes are made to the ligands (3).

The feverish interest in hydrido complexes has, as its main cause, the tremendous potential of these reactions in catalytic systems. In a relatively short span of time, hydrido complexes have been found to play a role in a significant number of catalytic processes (6, 9)—e.g., oligomerization of olefins (rhodium), decarboxylation reactions (rhodium), and hydrogenation reactions (ruthenium, osmium, rhodium, iridium, and platinum). Discussion of these applications would go beyond the scope of the present treatment.

The remainder of this review is devoted to giving a closer look at the hydrido complexes of iridium and their reactions. Iridium gives the most extensive and versatile range of hydrido complexes of any platinum metal (29). The main types of compounds known and some of their reactions are shown in Table II. To this, one must add the hydrogen abstraction reaction mentioned earlier.

The wide range of hydrido complexes formed by iridium is a clear indication that this metal has a particularly strong tendency to form Ir–H bonds. The reason for this affinity is not apparent as yet, but this conclusion is reinforced constantly by new data. Thus, one can isolate stable seven-coordinate complexes of iridium(III). The quadridentate ligand (o-Ph$_2$P · C$_6$H$_4$)$_3$P, QP, reacts with [IrHX$_2$(Ph$_3$P)$_3$] (X = Cl, Br, and I) in refluxing chlorobenzene to give [IrHX(Ph$_3$P)(QP)]X, I, which is (21) a seven-coordinate hydrido complex of iridium(III). The infrared spectrum does not show an Ir–H stretching vibration, but the presence of a hydrido ligand is clearly visible in the proton magnetic resonance spectrum.

The formation of these compounds is preceded by that of a species of composition [Ir(Ph$_3$P)(QP)]X, II, which reacts with Na[BPh$_4$] to give the same complex irrespective of the nature of the anion X in the starting material, [IrHX$_2$(Ph$_3$P)$_3$]. Their NMR spectra (22) indicate the presence of Ir–H bonds and their reaction with HX gives compounds [IrHX(Ph$_3$P)(QP)]X. Complexes II are, therefore, tentatively formulated as [IrH(o-C$_6$H$_4$PPh$_2$)(QP)]X.

In an attempt to gain a better understanding of the factors causing the formation of the above seven-coordinate hydrido complexes, the reaction of [IrHCl$_2$(Ph$_3$P)$_3$] with refluxing chlorobenzene was examined. Preliminary experiments suggest that the primary step is the evolution of hydrogen chloride and formation of the unstable intermediate [IrCl(Ph$_3$P)$_3$], which reacts further to give a number of products. One has the composition IrCl(Ph$_3$P)$_3$ and is likely to be an isomeric form, III, of Bennett and Milner's compound (7), [IrHCl(o-C$_6$H$_4$PPh$_2$)(Ph$_3$P)$_2$]. The others, some purely organic and some iridium-containing,

Table II. Some Iridium Hydrido Complexes and Their Reactions

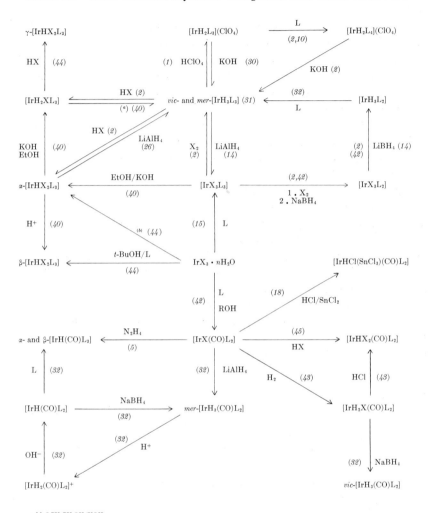

^a MeOCH$_2$CH$_2$OH/KOH
^b MeOCH$_2$CH$_2$OH/H$_2$O/L

are still under investigation (33). Thus, one can presume that compound III reacts with the ligand QP to give compounds of type I, accounting for the retention of the elements of one molecule of triphenylphosphine in compounds I. The interest in complexes I arises from the fact that they exhibit coordination number seven despite the d^6 electron configuration of the metal ion. The formation of such seven-coordinate species may be assisted by the fact that the seventh ligand is a hydrogen atom, but the significance of their isolation is that it is reasonable to postulate, when necessary, the formation of a seven-coordinate transition state intermediate even when the formal electron configuration of the metal atom is d^6.

The formation of such a range of hydridic compounds by iridium and the paucity of catalytic processes based on complexes of this element may be linked to the greater inertness of compounds of elements of the third transition series relative to the corresponding species with metals of the second transition series. However, iridium chemistry may provide a very fertile field for the study of homogeneous catalysis. The key to success in this direction may lie in the use of complexes which are coordinatively unsaturated and contain a strong labilizing group (6, 9). Thus, [IrHCl$_2$(Ph$_3$P)$_3$] in chlorobenzene solution reacts with dimethylformamide at 136° with formation of [IrCl(CO)(Ph$_3$P)$_2$], IV, which is in equilibrium with [IrHCl$_2$(CO)(Ph$_3$P)$_2$]. Compound IV catalytically decomposes dimethylformamide into carbon monoxide and dimethylamine, but the decomposition rate is proportional to the concentration of IV and amide, indicating that the rate-determining step is the formation of the adduct between IV and dimethylformamide and not the decomposition of the adduct which regenerates IV. When [IrHCl$_2$(Ph$_3$P)$_3$] reacts with dimethylacetamide, however, only a stoichiometric reaction, which is still under investigation, is observed (33).

This review has been necessarily brief, but I hope sufficient to show that the field of platinum metal hydrides is one of unlimited potential for both the academic and industrial chemist.

Literature Cited

(1) Angoletta, M., *Gazz. Chim. Ital.* **1962**, 92, 811.
(2) Angoletta, M., Araneo, A., *Gazz. Chim. Ital.* **1963**, 93, 1343.
(3) Atkins, P. W., Green, J. C., Green, M. L. H., *J. Chem. Soc.* **1968**, (A), 2275.
(4) Basch, H., Ginsburg, A. P., *J. Phys. Chem.* **1968**, 73, 854.
(5) Bath, S. S., Vaska, L., *J. Am. Chem. Soc.* **1963**, 85, 3500.
(6) Bird, C. W., "Transition Metal Intermediates in Organic Synthesis," Logos Press and Academic Press, London, 1967.
(7) Bennett, M. J., Milner, D. L., *J. Am. Chem. Soc.* **1969**, 91, 6983.
(8) Bresadola, S., Rigo, P., Turco, A., *Chem. Commun.* **1968**, 1205.
(9) Candin, J. P., Taylor, K. A., Thompson, D. T., "Reactions of Transition Metal Complexes," Elsevier, Amsterdam, 1968.
(10) Canziani, F., Zingales, F., *Rend. Ist. Lombardo Sci.*, A. **1962**, 96, 513.
(11) Chatt, J., *Science* **1968**, 160, 3829.
(12) Chatt, J., Butter, S. A., *Chem. Commun.* **1967**, 501.
(13) Chatt, J., Coffey, R. S., Shaw, B. L., *J. Chem. Soc.* **1965**, 7391.
(14) Chatt, J., Davidson, J. M., *J. Chem. Soc.* **1965**, 843.
(15) Chatt, J., Field, A. E., Shaw, B. L., *J. Chem. Soc.* **1961**, 290.
(16) Chatt, J., Hayter, R. G., *J. Chem. Soc.* **1961**, 2605.
(17) Chatt, J., Shaw, B. L., *J. Chem. Soc.* **1962**, 5075.
(18) Craig Taylor, R., Young, J. F., Wilkinson, G., *Inorg. Chem.* **1966**, 5, 20.
(19) Cross, R. J., Glockling, F., *J. Chem. Soc.* **1965**, 5422.

(20) Davison, A., McFarlane, W., Pratt, L., Wilkinson, G., *J. Chem. Soc.* **1962**, 3653.
(21) Dawson, J. W., Kerfoot, D. G. E., Preti, C., Venanzi, L. M., *Chem. Commun.* **1968**, 1689.
(22) Dawson, J. W., Venanzi, L. M., unpublished observations.
(23) Ginsberg, A. P., *Transition Metal Chem.* **1965**, 1, 111.
(24) Green, M. L. H., Jones, D. J., *Advan. Inorg. Chem. Radiochem.* **1965**, 7, 115.
(25) Griffith, W. P., "The Chemistry of the Rarer Platinum Metals," Interscience, London, 1967.
(26) Hayter, R. G., *J. Am. Chem. Soc.* **1961**, 83, 1259.
(27) Kingston, J. V., Scollary, G. R., *Chem. Commun.* **1969**, 455.
(28) Levison, J. J., Robinson, S. D., *Chem. Commun.* **1968**, 1405.
(29) Malatesta, L., *Helv. Chim. Acta*, Fasciculus Extraordinarius **Alfred Werner, 1967**, 147.
(30) Malatesta, L., "Symposium on Current Trends in Organometallic Chemistry," (Abstracts), Cincinnati, June 12-15, 1963, p. 71A.
(31) Malatesta, L., Angoletta, M., Araneo, A., Canziani, F., *Angew. Chem.* **1961**, 73, 273.
(32) Malatesta, L., Caglio, G., Angoletta, M., *J. Chem. Soc.* **1965**, 6974.
(33) Martelli, M., Venanzi, L. M., unpublished observations.
(34) Misono, A., Uchida, Y., Hidai, M., Araki, M., *Chem. Commun.* **1968**, 1044.
(35) Morelli, D., Segre, A., Ugo, R., LaMonica, G., Cenini, S., Conti, F., Bonati, F., *Chem. Commun.* **1967**, 524.
(36) Pidcock, A., Richards, R. D., Venanzi, L. M., *J. Chem. Soc., A* **1966**, 1707.
(37) Sacco, A., Rossi, M., *Inorg. Chim. Acta* **1968**, 2, 127.
(38) Sacco, A., Ugo, R., *J. Chem. Soc.* **1964**, 3274.
(39) Shaw, B. L., "Inorganic Hydrides," Pergamon, Oxford, 1967.
(40) Shaw, B. L., Chatt, J., *Proc. Intern. Congr. Coordination Chem., 7th, Stockholm, 1962*, Abstracts, p. 293.
(41) Vaska, L., DiLuzio, J. W., *J. Am. Chem. Soc.* **1961**, 83, 1262.
(42) Vaska, L., DiLuzio, J. W., *Ibid.*, **1961**, 83, 2784.
(43) Vaska, L., DiLuzio, J. W., *Ibid.*, **1962**, 84, 679.
(44) Vaska, L., DiLuzio, J. W., *Ibid.*, **1962**, 84, 4989.
(45) Vaska, L., Rhodes, L. E., *J. Am. Chem. Soc.* **1965**, 87, 4970.

RECEIVED January 28, 1970.

7

Ultraviolet Spectroscopic Qualities of Platinum Group Compounds

DON S. MARTIN, JR.

Institute for Atomic Research and Department of Chemistry, Iowa State University, Ames, Iowa 50010

> *Coordination compounds of the platinum group elements have excited states which provide spectral absorptions in the ultraviolet region, extending through the visible into the near IR, which must be classified together from their similarity in origin. Many of the transitions are obscured by the broad bands from their more intense neighboring transitions, so only a few can be identified for a single complex. Several analytical and kinetics applications are discussed. In addition, applications of polarized crystal spectra, absorption in low-temperature organic glasses, and the use of magnetic circular dichroism which can be combined with molecular orbital treatments, spin-orbit coupling, vibronic excitation, and metal–metal interactions are described for the identification of transitions in specific instances.*

The bright colors of the coordination complexes of transition metal elements, including the platinum group metals, were of great assistance to pioneer workers with these materials. Thus, chemical changes could be followed visually; it was frequently very easy from their colors to demonstrate the existence of isomers upon which Alfred Werner was able to base his monumental theory of coordination. Such early studies were limited to a simple qualitative visual evaluation of the color.

By the mid-twentieth century, dependable spectrophotometers became commercially available and very quickly were recognized as essential laboratory tools for the characterization of the coordination complexes. These new instruments permitted quantitative evaluation of the absorption of light; procurement of the instruments was usually justified by the convenience, versatility, and speed which they introduced into the quantitative analysis and identification of these materials.

Spectrophotometry is today, without question, the most widely used instrumental technique for kinetics studies, which require the quantitative characterization of changing systems.

Modern instruments have extended the spectral range from the severely limited visual region into both the near infrared and the ultraviolet region. A region extending from 1200 down to 200 mμ is rich with the electronic spectra involving the d-orbitals of transition metal compounds. Since strict adherence to the ultraviolet region would be so incomplete, consideration here will be directed to the broader region of the accessible electronic spectra for the platinum metal complexes.

The literature contains literally hundreds of absorption spectra for the coordination complexes of the platinum elements, spectra which depend upon the nature of the various ligands, the oxidation state, and the electronic configuration. In the present review, a complete cataloging of all of these spectra will not be attempted. Rather, attention will be limited to a few complexes for which the ligands are relatively simple groups, such as halides, ammonia, water, or simple amines which by themselves do not possess a high absorption in the spectral region of interest. Typical illustrations of the analytical possibilities as well as means of chemical characterization with the systems are presented. However, the analytical utility of the spectra, although quite impressive, has been overshadowed within the past 15 years by the application of crystal or ligand field theory, and subsequently by molecular orbital treatments, which have provided an understanding of the observed absorption bands and their assignment to predictable electronic transitions. The development of these theories has been one of the most exciting events in modern inorganic chemistry, for they have provided a means of identifying the electronic states and testing various concepts and theoretical calculations of bonding.

One of the very active regions of research in recent years has involved attempts to extend experimental methods to provide additional spectral information for more definitive electronic state assignments.

Spectra of Hexahalo-Complexes in Aqueous Solutions

The platinum metals in the heavy series form especially inert hexahalo-complexes. Spectra for a number of these were reported by Jorgensen (*14, 15, 16*). A few of these which are illustrative of the effects of various ligands and of the electronic configurations have been reproduced in Figure 1. Usually, the solutions from which these spectra were obtained have contained excess halide ion to reduce any solvation effect. A number of limitations for the analytical and structural utility of the spectra are immediately evident from the spectra in Figure 1. First of

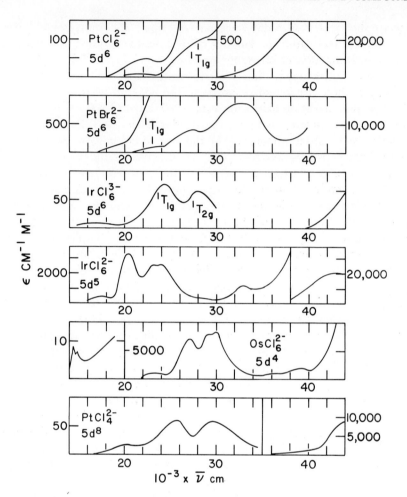

Figure 1. Electronic absorption spectra for some halide complexes of the heavy platinum group elements

all, the spectra consist of a limited number of very broad absorption bands. It is not possible to determine from the spectrum of a mixture of several complexes the composition of the mixture, or frequently even to identify the components. Nevertheless, some conclusions of considerable value are possible.

For $PtCl_6^{2-}$, the molar absorptivity at 38,000 cm^{-1} is $> 20,000$ cm^{-1} M^{-1}. For a pure solution, a 10-cm cell will yield an absorbance of 0.1 for a 0.5 μM solution. High sensitivity in analysis is therefore possible, provided interferences are absent. The molar absorptivity for $PtCl_6^{2-}$ at 30,000 cm^{-1} is 600 cm^{-1} M^{-1}, whereas the molar absorptivity for $PtBr_6^{2-}$ is 10,000 cm^{-1} M^{-1}. It is possible from the spectra, therefore, to place a fairly small limit upon the contamination by the $PtBr_6^{2-}$ which may be present. Whereas the height of a spectral peak is a good measure for the concentration of the predominant species, the absorbance at a spectral valley is very sensitive to the presence of impurities. For synthetic preparations, one of the most reliable criteria of purity is the peak-to-valley ratio in an absorption spectrum. A very valuable preparative technique involves continued fractional recrystallization until such a peak-to-valley ratio does not increase upon each subsequent recrystallization.

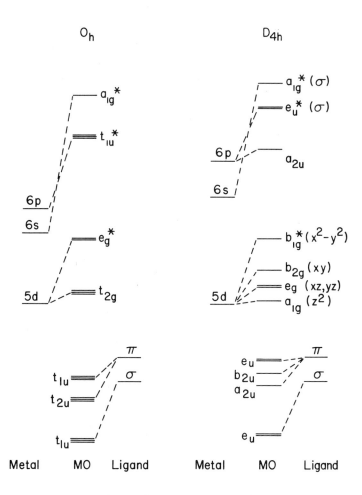

Figure 2. Correlation diagram for molecular orbitals in coordination complexes with O_h symmmetry and with D_{4h} symmetry. Only the principal correlation lines are shown and the g states arising from the ligand σ and π orbitals are omitted for simplicity

Only a handful of broad electronic transition bands, several thousand wave numbers in width, can be identified from the spectrum of any one complex. Frequently, not more than six such bands are resolvable. Jorgensen has provided interpretations of these spectra (*17*, *18*) which account for many of their observed features. A 1-electron molecular orbital correlation diagram for the O_h symmetry group, which applies to these systems, has been included in Figure 2. A sigma orbital from each of the six ligands interacts with the s, the p, and the d-(e_g) orbitals to provide six sigma bonding orbitals. Only the three asymmetric t_{1u} orbitals of this set are indicated in Figure 2. The two π orbitals of each ligand generally are not considered to provide a strong contribution to the bonding for halides. Two sets of three-fold degenerate orbitals, the t_{1u} and the t_{2u}, are shown. The levels for the symmetric (g) sigma and π orbitals have been omitted for simplicity.

A set of symmetric orbitals, the t_{2g} comprising principally the metal d-orbitals, are next in energy and above them, separated by the familiar Δ_o or 10 Dq, are the antibonding e_g^* pair. At much higher energies are the antibonding t_{1u}^* and the a_{1g}^* orbitals. Since for the heavy series of platinum metals Δ_o is of the order of 30,000 cm^{-1}, the complexes are of the spin-paired and inert type. Up to six electrons can be in the t_{2g} set of orbitals, and all orbitals below these are filled as well in the complexes. A transition of an electron from the filled orbitals, t_{1u} (π) or t_{2u} (π), to either the t_{2g} or the e_g^* orbital will be dipole-allowed. Since in these transitions the electron moves primarily from an orbital on the ligand to one on the metal, such a transition is usually designated as a $L \rightarrow M$ charge transfer. Such transitions are expected to lead to bands with the molar absorptivity, ϵ, at the peak greater than 1000 cm^{-1} M^{-1}. Transitions of electrons from t_{2g} to t_{1u}^* would be allowed by symmetry, and the possibility of their presence in the spectra must be considered as well. However, in most instances they are believed to occur at higher energy, frequently beyond the range of observations with halides.

Transitions of an electron from a t_{2g} orbital to e_g^* are dipole-forbidden. Bands with a molar absorptivity of the order of 100 cm^{-1} M^{-1} usually are assigned to such d-d transitions in which the spins are unchanged—*i.e.*, they are spin-allowed. Since the heavy metals have a high spin-orbit coupling, and the spin is no longer a good quantum number by the mixing of different multiplet states, bands are observable between states which have major components with different spin multiplicities. These are designated as the spin-forbidden bands. Indeed, the mixing of states is sufficiently great that there may be considerable ambiguity in the assignment of transitions. The situation is treated by considering the double rotational groups which encompass both the orbital and spin wave functions. For example, the electron spin functions, $+1/2$, $-1/2$, are basic functions for the $\gamma_6(e_2)$ irreducible representation for the O' group. Singlet and triplet functions have $\gamma_1(a_1)$ and $\gamma_4(t_1)$ symmetry, respectively. The actual states may be formed from a linear combination of wave functions with different multiplicities. The contribution of minor multiplicities may be sufficiently great that "spin-allowed" transitions have an enhancement factor of only 4 to 10 over "spin-forbidden" transitions. As the wavelength decreases, it is frequently possible to observe in a spectrum first some spin-forbidden, symmetry-forbidden transitions, then the spin-allowed but symmetry-forbidden, and finally some charge transfer bands. However, only a fraction of the possible transitions are observable, and many are "hidden" under the broad peaks of more intense transitions in solution spectra of the type illustrated in Figure 1.

Theory has not yet reached a stage where energies or even the ordering of states can be calculated reliably for transition element complexes. Therefore, complementary contributions and cooperation between experiment and theory are needed in the unravelling of energy states and the description of bonding for these important systems.

The d^6-configuration, which occurs for the multitude of inert Co(III) complexes, is illustrated by the complexes of Ir(III) and the Pt(IV) complexes in Figure 1. The ground state configuration is primarily $t^6_{2g} - {}^1A_{1g}$.

Since the t_{2g} orbitals are filled, there is no possibility of a transition from the ligand π-orbital into one of these; therefore, charge transfer bands will involve only transition to the antibonding $e_g{}^*$. The strong band peaking at 38,000 cm^{-1} with ϵ of 24,000 cm^{-1} M^{-1} for PtCl$_6{}^{2-}$ was attributed to such transition by Jorgensen. The transition $\pi(t_{1u}) \rightarrow e_g{}^*$ leads to four energy levels whose principal components are ${}^1T_{1u}$, ${}^1T_{2u}$, ${}^3T_{1u}$, ${}^3T_{2u}$. Because of the t_1 character of the triplet spin functions, transitions to the primarily ${}^1T_{1u}$, ${}^3T_{1u}$, ${}^3T_{2u}$ states all have some dipole-allowed character. However, no resolution into such individual states is possible with the observed band nor can any decision be presented as to whether a contribution from $\pi(t_{2u}) \rightarrow e_g$ occurs as well. There is the possibility of a symmetry-forbidden transition of one of the t_{2g} electrons into the $e_g{}^*$. The excited electron in either of two orbitals and the residual hole in any of three leads to a total of six singlet and six triplet states. However, these six states in each case constitute two three-fold degenerate sets; the lower in energy can be designated as T_{1g} and a higher T_{2g}. Two spin-allowed bands should occur, therefore. As indicated with Jorgensen's assignment in Figure 1, only the first transition can be seen as a shoulder on the much more intense charge transfer band. The still lower intensities in the region of 19,000 to 24,000 cm^{-1} will correspond then to the spin-forbidden symmetry-forbidden bands.

For the hexabromoplatinate(IV), there has been a decided shift of all bands to lower energies or longer wavelengths, and the breadth of the charge transfer band definitely suggests more than a single excited state. This energy shift between chloride and bromide ligands is quite a general phenomenon representing the position of these ligands in the spectrochemical (sometimes called the Fajans-Tsuchida) series. The shift of the charge transfer band reflects the easier oxidation or loss of an electron by bromide in comparison with chloride. However, the shift in the d-d transition represents the difference in the crystal field strength of these two ligands. The $IrCl_6^{3-}$ ion, with the same electron configuration as the two platinum complexes, however, is in a lower oxidation state. Additional energy is required to add an electron to the iridium, and consequently the charge transfer band has been forced to much higher energies. Here, the transitions to the two excited states, primarily $^1T_{1g}$ and $^1T_{2g}$, are clearly discernable. The occurrence of the d-d bands at lower energies corresponds to a lower Δ_o for Ir(III) than for Pt(IV), which is in accordance with the usual systematics in this quantity (19). The breadth of the high-intensity band, 31,000–32,000 cm^{-1}, probably reflects the higher spin-orbit coupling of the bromide ligands in comparison with that of chloride.

For Ir(IV) in $IrCl_6^{2-}$, there is a hole in the t_{2g} orbital set for the ground state, $t_{2g}^5 - {}^2T_{2g}$. Therefore, the transition of an electron from one of the ligand π-orbitals into a t_{2g} orbital is possible. Jorgensen (17) proposed that these charge transfer bands were the ones seen in the visible region, 20,000–25,000 cm^{-1}, which had ϵ's of 2000–4000 cm^{-1} M^{-1}. These bands provide the intense color for $IrCl_6^{2-}$ so that its presence in minute amounts can be detected easily in solutions of $IrCl_6^{3-}$. The presence of hyperfine structure on the electron spin resonance spectrum of this (8, 25) ion indicated that the t_{2g} orbitals were not solely metal orbitals, but that they contained significant ligand orbital character. This feature demonstrated the existence of some π-type bonding in the complex.

For Os(IV), there are two electrons missing from a filled set of t_{2g} orbitals. Therefore, transition from the ligand π-orbitals to the t_{2g} orbitals should again be possible. Since Os(IV) is a much weaker oxidizing agent than Ir(III), it is expected that these charge transfer bands will be shifted to higher energies. The spectrum in Figure 1 indicates that an absorption pattern very similar to that for $IrCl_6^{2-}$ occurs for $OsCl_6^{2-}$. Accordingly, Jorgensen assigned the bands from 28,500–30,000 cm^{-1} to these $L \rightarrow t_{2g}$ transitions. The $OsCl_6^{2-}$ spectrum is one of the richest to be observed. Still, only a handful of incompletely resolved broad bands in the solution spectrum are seen. Since a considerably larger number of such charge transfer states may be attained by dipole-allowed transitions than are observed, a detailed confirmation of this assignment has not been possible. Some very recent information, which will be discussed later, has brought this assignment of these bands into question. Figure 1 does show two very weak, relatively sharp bands that occur at approximately 17,000 cm^{-1} for OsCl$_6^{2-}$ which are not observed in the other spectra presented. These are believed to represent a transition within the ground state configuration—i.e., $t^4{}_{2g}\Gamma_1(^3T_{1g}) \rightarrow \Gamma_1(^1A_{1g})$. Still weaker transitions (12) have been observed at about 8000, 10,800, and 11,780 cm^{-1}.

Many features of the spectra of the octahedral complexes have seemed consistent with simple molecular orbital models. The high symmetry possessed by these complexes fortunately has reduced the number of individual transitions which should occur because of the many degenerate states for the ion. The broadness of the few bands observed, however, precludes resolution of the fairly large number of transitions which might be expected to have measurable intensities as a consequence of spin-orbit coupling. The breadth of the absorption bands at least partially arises from thermally excited fluctuations in the environment of the coordination complex. For example, in the ground state a bond may be lengthened by a molecular vibration. The vibrational energy in this mode will amount to only about kT. However, because of the Frank-Condon principle, the excited ion will have the same inter-atomic distances as the ground state. Thus, if the excited state ion has a steep potential energy function for atomic separations at the transition configuration, a vibrational displacement of modest energy in the ground state will produce large energy differences in the excited complex. A reduction in temperature will lead accordingly to much narrower and better resolved bands.

Spectrum of PtCl$_4^{2-}$

The PtCl$_4^{2-}$ ion, with the 2+ oxidation state for platinum, possesses a $5d^8$ configuration. It has a square-planar coordination, which is commonly observed with the low spin complexes of this configuration. Its absorption spectrum in aqueous solution is also shown in Figure 1. With the platinum in a low oxidation state, the charge transfer bands have been shifted to beyond 40,000 cm^{-1}.

Since PtCl$_6^{2-}$ absorbs so strongly at 38,000 cm^{-1}, the absorbance at this wave number can be used to set a very low limit of detection on the presence of Pt(IV) in samples of PtCl$_4^{2-}$. This is especially valuable, since the higher oxidation state has sometimes been suspected as a catalyst for some of its reactions.

The molecular orbital scheme for the D_{4h} symmetry, possessed by this ion, is shown in Figure 2. Although the ion is highly symmetric, the symmetry is sufficiently low that information from polarized spectra of crystals is very useful in establishing assignments for the transitions. The peaks at 25,400 and 30,300 cm^{-1}, with ϵ about 60, are clearly spin-allowed d-d transitions. It was not clear whether the smaller peak at about 20,000 cm^{-1} corresponds similarly to a spin-allowed transition or to one which is spin-forbidden. Figure 2 shows a possibility of three spin-allowed transitions from the filled d-orbitals into the anti-bonding or the $d_{x^2-y^2}$ orbital. In such a transition, the electron leaves a symmetric (g) orbital and moves into a symmetric (g) orbital, so the transition is therefore dipole-forbidden. However, these transitions become allowed in the presence of an asymmetric field. Such a field may result from the chemical environment—i.e., the effect of neighboring molecules or ions in a solution or in a crystal. It can also result from the asymmetric vibrations of the complex. In the latter case, the transition is said to be vibronically excited. In quantum mechanical description, the vibration is considered a perturbation on the system. Thus, the electronic wave function, ψ_{elect}, is represented by the equation

$$\psi_{\text{elect}} = \psi_g - \sum_i \sum_{\psi_u} CQ_i \psi_u \qquad (1)$$

where ψ_g is the unperturbed function, ψ_u is an asymmetric wave function, Q_i is as an asymmetric normal vibration, and

$$C = \int \psi_g^* \left(\frac{\partial H}{\partial Q_i}\right) \psi_u \, d\tau / E_u - E_g \qquad (2)$$

It is necessary to analyze the symmetry properties of the normal vibrations of the complex in order to obtain a set of selection rules. Therefore, the spin-allowed transition of an electron from the d_{xy} to the $d_{x^2-y^2}$ orbital will occur with the electric vector of light polarized in the direction normal to the symmetry axis—i.e., in the xy plane. Transitions taking an electron from either the d_{z^2} or $d_{xz,yz}$ orbitals to the $d_{x^2-y^2}$ will occur with polarization in the z direction and the xy direction as well. K$_2$PtCl$_4$ provides ideal crystals to test these conclusions. In the first place, each PtCl$_4^{2-}$ occupies a site with full D_{4h} symmetry (6), so no asymmetries can be introduced from the crystal fields. Each ion is aligned with its symmetry (z) axis directed along the tetragonal c-axis. Crystals were grown as thin plates, $< 50\mu$ thick, with the c-axis lying in the face of the plates. It was possible to measure the absorption of light polarized in the z direction and normal to it (20). These spectra are shown in Figure 3, both for room temperature and with liquid helium cooling, nominally at 15°K. A band at ca. 26,000 cm^{-1} in xy polarization is completely missing

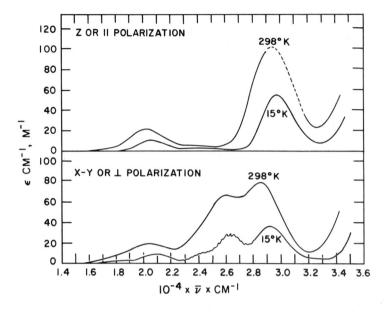

Figure 3. Absorption spectra of a K_2PtCl_4 crystal with polarized light

Crystal thickness, 46 microns

with the z polarization. This appears to be a rigorous selection rule and identifies that band as associated with the $d_{xy} \rightarrow d_{x^2-y^2}$ transition. Incidentally, the crystal spectra identified the band at 20,000 cm^{-1} as due to spin-forbidden transitions. If it were a spin-allowed transition, then the theory would require an observable spin-forbidden band in the region of 11,000–14,000 cm^{-1}. No such absorption was observed in this region, even for crystals 1 mm thick.

Assignment of the band at about 30,000 cm^{-1} was achieved from a measurement of the magnetic circular dichroism (MCD). When a species with a nondegenerate ground state absorbs light to form a degenerate excited state, the degeneracy will be broken by the Zeeman splitting. A characteristic MCD spectrum results with an inversion in sign of the measured $A_R - A_L$ occurring near the absorbance maximum. This has been assigned the A-term in the theoretical treatment of Stephens (26). The MCD spectrum in Figure 4 shows just this characteristic form associated with the band at 30,000 cm^{-1}. Hence, it can be assigned to the transitions from the degenerate pair of orbitals $d_{xz,yz} \rightarrow d_{x^2-y^2}$.

At low temperatures, narrower bands with better resolution are expected, and the vibronic model predicts lower intensities as well. The polarized spectra at liquid helium temperatures confirm these predictions. Thus, shoulders at 24,000 cm^{-1} and in the region of 17,000–19,000 cm^{-1}

which could not be resolved at room temperature now are resolved clearly. An interesting feature is the fine structure which appeared on the two bands from 23,000–28,000 cm^{-1}, where 17 individual peaks can be seen. These individual peaks result from excitation of the totally symmetric, breathing vibration in the excited electronic states. Weaker vibrational structure could be discerned, especially in z polarization for the band centered on 20,600 cm^{-1}. However, no such structure was evident on the strong band at 29,600 cm^{-1}.

An interesting conjecture is that bands for which the vibrational structure is missing involve transitions to excited electronic states which

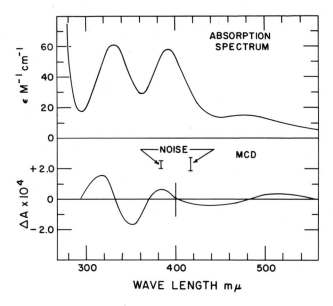

Figure 4. Absorption spectrum and magnetic circular dichroism for a solution of K_2PtCl_4 in dilute aqueous HCl
Cell length, 1.00 cm
Concn. K_2PtCl_4, 0.015M

correlate with states of lower energy in tetrahedral coordination of the ligands. One of the out-of-plane bending vibrations carries the square-planar coordination toward the tetrahedral. Consequently, a harmonic potential energy function for this vibration would not apply. Such a consequence would likely provide many more possible vibrational states in the excited state. The abundance of closely spaced states which are then possible prevents their resolution by the technique employed.

The energy of the transition from the d_{z^2} to the $d_{x^2-y^2}*$ is not indicated from the observed spectra. However, the 1-electron orbital energies calculated from theory by Basch and Gray (2) and by Cotton and Harris (4) place the d_{z^2} orbital well below the degenerate d_{xz}, d_{yz} orbitals. Therefore, they have assigned absorption in the region of 38,000–40,000 cm^{-1} on the tail of charge transfer bands to this transition.

The energy states for a d^8 system can be calculated from a set of parameters including the splitting of the d-orbitals, the electron–electron repulsion parameters (either those of Slater and Condon or of Racah) and the spin-orbit coupling. In Figure 5 are presented the results of such a calculation, with the parameters adjusted to assign transitions with expected high intensity to the observed bands. The results of a corresponding calculation for the tetrahedral arrangement are shown also.

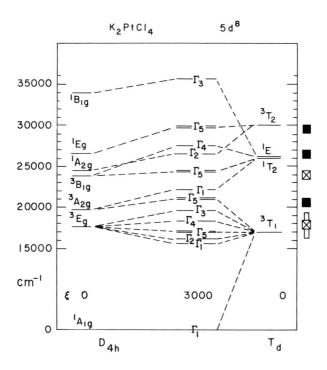

Figure 5. Calculated energy levels for the $PtCl_4^{2-}$ ion in D_{4h} and T_d symmetry. In the center, the energy levels are shown for D_{4h} coordination with the inclusion of a spin-orbit coupling parameter: $\zeta = 3000$ cm^{-1}. The states are designated with the Γ notation for the D_4' double group, which applies with the inclusion of spin-orbit coupling. For D_{4h}: $d_{x^2-y^2} - d_{xy} = 26,100$ cm^{-1}; $d_{x^2-y^2} - d_{xz,yz} = 30,200$ cm^{-1}; $d_{x^2-y^2} - d_{z^2} = 38,300$ cm^{-1}, $F_2 = 1000$ cm^{-1}, $F_4 = 65$ cm^{-1}. For the T_d coordination, $\Delta_t = 14,000$ cm^{-1}, $F_2 = 1000$ cm^{-1}, and $F_4 = 65$ cm^{-1}. Observed bands for the crystal are shown at the right side

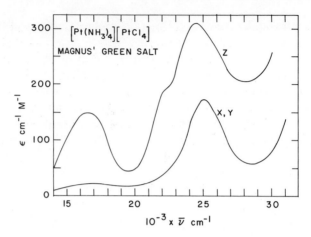

Figure 6. Absorption spectrum of Magnus' Green Salt, $Pt(NH_3)_4PtCl_4$, with polarized light

It is possible to select a value of the parameter Δ_t so that with the indicated values of F_2 and F_4 the bands with high vibrational structure at low temperature correlate with much higher energy states in the tetrahedral symmetry. This figure illustrates the point that spin-orbit coupling multiplies the number of separated energy states. Even with large spin-orbit couplings of the heavy platinum elements, which provide the intensity for these spin-forbidden transitions, the splitting of the many states is not sufficient for their individual resolution from the experimental solution spectra where only a few broad areas of absorption can be identified.

Professor Schatz and coworkers (23) at the University of Virginia have extended the MCD studies into the charge transfer band region. Here also, they identify a degeneracy for the excited states, which presumably arises from a transitional electron leaving an $e_u(\pi)$ orbital. However, they concluded that the strongest observed band at 46,200 cm^{-1} corresponded to a $5d \rightarrow 6p$ transition.

The absorption bands for the crystals of K_2PtCl_4 resemble those of the solutions in both intensity and energy. There is little doubt that for the crystal the observed spectra are caused by isolated ions which are not influenced substantially by neighboring ions. In these crystals, the planar ions are stacked one above the other with a separation of 4.21A. The potassium ions pack in the planes between the ions with contacts to the external portion of the chloride ligands. In the compound $Pt(NH_3)_4PtCl_4$, Magnus' green salt (MGS), there are striking differences in the spectra. In MGS, the cation and anions stack alternately with a separation of only 3.25A along a needle axis of the crystal. These needles are strongly dichroic, with a maximum absorption in the visible in the direction of the stacked chains. Polarized spectra, which were reported by Day et al. (5), are shown in Figure 6. Since absorption bands of the ion Pt(NH$_3$)$_4^{2+}$ occur at higher energies than for PtCl$_4^{2-}$, the spectrum of MGS is believed to represent that of the PtCl$_4^{2-}$ ion as modified by the close presence of the cations. The high absorption in the z direction is evident from Figure 6. However, the intense green results from a rather normal absorption band at 16,500 cm^{-1} with a window at 20,000 cm^{-1} in the spectrum for a coordination complex. There is no evidence for an absorption continuum characteristic of a conduction type band. However, the high absorption in the z direction has been considered a consequence of a perturbation of the states for PtCl$_4^{2-}$. A z-polarized, dipole-allowed transition is moved to lower energies closer to the *d-d* transitions. Consequently, greater intensity for z-polarization can be borrowed under the vibronic mechanism (*1, 5, 24*).

Effects of Reduction in Symmetry

If one of the chloride ligands in PtCl$_4^{2-}$ is replaced by another group, the center of symmetry in the ion is removed. In Figure 7 are shown spectra in the region of 25,000–40,000 cm^{-1} for the ions PtCl$_3$(H$_2$O)$^-$ and PtCl$_3$(NH$_3$)$^-$. This is the region for the spin-allowed *d-d* bands of these ions. In accordance with the concept that ammonia has a stronger crystal field effect than H$_2$O, which is stronger than that of chloride, bands are shifted to higher energy by the substitution of a chloride by water and to still higher energy upon substitution by ammonia. Since there is no center of symmetry in these ions, the presence of the ligands creates an asymmetric electric field at the platinum atom. The higher intensities of the bands seem to be a natural consequence of these asymmetric fields. One puzzling feature for which there appears to be no ready explanation is the relative intensities of the two bands in the three complexes. For PtCl$_4^{2-}$, the ϵ's of the two spin-allowed peaks are about equal. In the PtCl$_3$(H$_2$O)$^-$, the high energy band is much more intense than the low energy band, whereas for the PtCl$_3$(NH$_3$)$^-$, just the reverse situation occurs.

If a sample of K$_2$PtCl$_4$ is dissolved in water with no additional chloride, and the absorbance of the solution is followed at 25,500 cm^{-1}—*i.e.*, at the maximum of the first spin-allowed band—the absorbance falls and the maximum shifts toward longer wavelengths as the spectrum changes continuously toward the spectrum of the PtCl$_3$(H$_2$O)$^-$. At equilibrium with respect to the aquation reaction

$$PtCl_4^{2-} + H_2O \rightarrow PtCl_3(H_2O)^- + Cl^- \qquad (3)$$

the spectrum is intermediate between the two shown in Figure 7. The spectrum for PtCl$_3$(H$_2$O)$^-$ was obtained from the observed equilibrium spectrum and the equilibrium concentrations of the species. The spec-

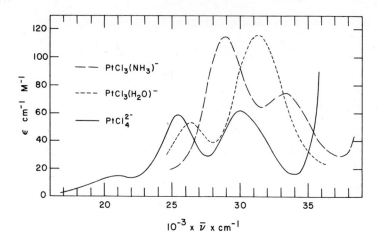

Figure 7. Absorption spectrum in dilute aqueous HCl of the complexes, $PtCl_4^{2-}$, $PtCl_3(H_2O)^-$, and $PtCl_3(NH_3)^-$

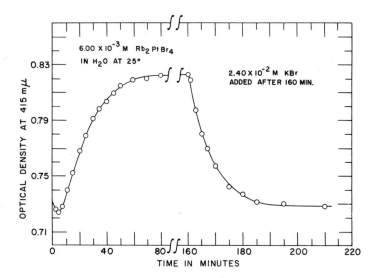

Figure 8. Changes in the absorbance after Rb_2PtBr_4 is dissolved in H_2O

Wavelength, 415 mµ. $NaNO_3$ added to give an ionic strength of 0.318M. H^+, 10^{-3}M.

trum of the $PtCl_4^{2-}$ ion can be evaluated fairly accurately in solutions with high chloride concentrations.

A similar effect was expected for a solution of $PtBr_4^{2-}$ at the corresponding band which occurs at 24,400 cm^{-1}. However, in such a solution, as shown in Figure 8, although there was initially a very small decrease in the absorbance, this decrease was followed by a much larger increase (27). Upon the addition of excess bromide ion, the absorbance returned to its initial value with a period of about 5 min. This behavior was attributed to the formation of the $Pt_2Br_6^{2-}$ dimeric ion in the solution. This ion had been identified previously by Harris et al. (9). The data in Figure 8 provided information about the rate of formation and reaction of this dimer. Apparently an aquation reaction

$$PtBr_4^{2-} + H_2O \rightarrow PtBr_3(H_2O)^- + Br^- \qquad (4)$$

occurs initially, but this process is followed by the reaction

$$2PtBr_3(H_2O)^- \rightarrow Pt_2Br_6^{2-} + 2H_2O \qquad (5)$$

This dimeric ion can be considered in equilibrium with $PtBr_4^{2-}$ in accordance with the reaction

$$2PtBr_4^{2-} \rightleftarrows Pt_2Br_6^{2-} + 2Br^- \qquad (6)$$

Since in the formation of $Pt_2Br_6^{2-}$ from $PtBr_4^{2-}$ two bromide ions are formed, there will be a higher fraction of the platinum converted to the dimer at higher dilution. Figure 9 illustrates the spectra recorded in an experiment in which a solution of $H_2Pt_2Br_6$ was diluted. The initial solution had been formed by passing the tetraethylammonium salt of $Pt_2Br_6^{2-}$ through a cation exchanger in the hydrogen phase. Upon dilution, the absorbance decreased toward a spectrum to be expected for the $PtBr_3(H_2O)^-$ ion. With the addition of excess KBr, the observed spectra coincided with that of $PtBr_4^{2-}$. The observed spectral changes indicated that the formation of the dimeric species, $Pt_2Br_6^{2-}$, was too slow to account for the isotopic exchange between free bromide ion and the bromide ligands in $PtBr_4^{2-}$, even though the kinetics for such exchange included a term which was second order in complex. It was concluded that such exchange occurred through the making and breaking of singly bromide-bridged dimeric complexes.

Spectra in Organic Media

Although the visible and ultraviolet spectra of the platinum metal complexes serve as a valuable means to follow their reactions, this feature is especially disturbing to one who is trying to establish the energy states for a particular species. Some care is necessary to establish that the measured spectrum results from the particular complex species of interest, and not some of its reaction products. Water is an effective nucleophile for these complexes. To avoid solvation of some platinum complexes, Mason and Gray (21) have recently reported spectra of the tetrabutyl-

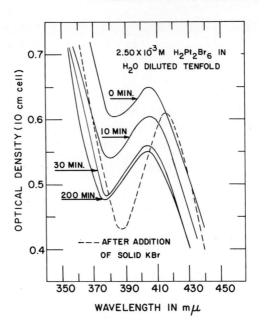

Figure 9. Spectral changes when an aged solution of $H_2Pt_2Br_6$ in a 1-cm cell (0 min) is diluted 10-fold and added to a 10-cm cell

Dashed curve obtained after the addition of solid KBr. Temp., 25°. $NaNO_3$ added to give ionic strength of 0.318M.

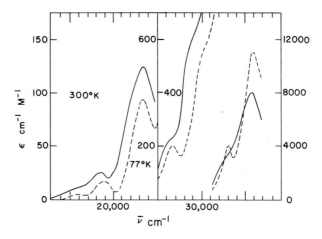

Figure 10. Spectrum of $(NBu_4)_2PtBr_4$ in a 2:1 mixture of 2-methyltetrahydrofuran and propionitrile at room temperature and at 77°K (liquid nitrogen)

ammonium tetrabromoplatinate(II) in an organic solvent medium consisting of 2:1 mixture of 2-methyltetrahydrofuran and propionitrile. In this way solvation was avoided. In addition, they were able to cool these solutions as glasses to liquid nitrogen temperatures. This provided much better resolution of the absorption bands, as is shown in their spectra in Figure 10. Shoulders at 26,500 and 33,000 cm^{-1} are clearly resolved at the lower temperatures. Since vibronic excitation is reduced at lower temperatures, intensities of the symmetry-forbidden bands in the region of 20,000–30,000 cm^{-1} are lower at the liquid nitrogen temperatures. The dipole-allowed transitions, on the other hand, possess a temperature-independent total intensity. Therefore, as the bands narrow at low temperature, extinction coefficient peaks become higher. This is especially

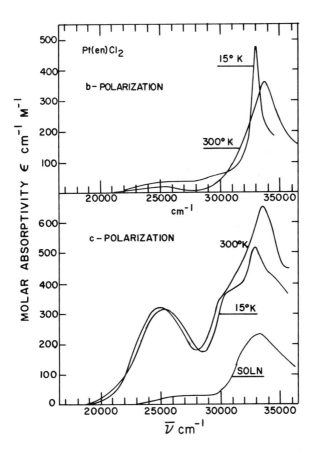

Figure 11. Absorption spectrum of Pt(en)Cl$_2$ in an aqueous 0.3M Cl$^-$ solution and in single crystals with polarized light at 300°K and at liquid helium temperatures (nominally 15°K)

evident in their spectra from 33,000–38,000 cm^{-1}, the region of the high intensity charge transfer bands.

Spectrum of Dichloro(ethylenediamine)platinum(II). Evidence for Metal–Metal Interactions

We have recently obtained some interesting spectra of the neutral complex Pt(en)Cl$_2$. The solution spectrum of this complex in the region of the d-d bands is shown in Figure 11. The charge transfer bands are beyond 48,000 cm^{-1}. Its solution spectrum is very similar to that of Pt(NH$_3$)$_2$Cl$_2$, except that the molar absorptivity in the region of the absorption maximum is twice as high for the cis-dichlorodiammineplatinum(II). Chatt, Gamlen, and Orgel (3) have assigned the spectrum of the dichlorodiammine complex on the basis of D_{4h} symmetry, considering the ligand field distortion to be a minor effect. The peak at 33,000 cm^{-1} was assigned as the first spin-allowed d-d band—i.e., $d_{xy} \rightarrow d^*_{x^2-y^2}$. The band for $d_{xz,yz} \rightarrow d^*_{x^2-y^2}$ was considered the shoulder to the high energy side of this peak. There were then two fairly well resolvable spin-forbidden bands at longer wavelengths. The crystal structure of Pt(en)Cl$_2$ was recently determined by x-ray diffraction methods by Jacobson and Benson in our laboratory (11). They found that the crystals belonged to the orthorhombic system with the molecules stacked in a nearly linear array, as shown in Figure 12, with a uniform spacing of 3.39A between the

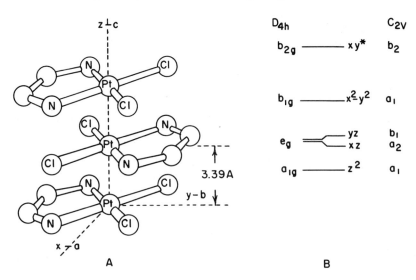

Figure 12. A. Stacking of Pt(en)Cl$_2$ molecules in the orthorhombic crystals. B. Anticipated order of 1-electron orbital energies with their symmetry designations for the d-orbitals under the D_{4h} and the C_{2v} groups.

platinum atoms. This spacing is only 0.14A greater than between the ions in MGS. Very thin crystals, $< 10\mu$ thick, with the c and b axes lying in the faces were available so that light directed along the a axis permitted the observation of spectra polarized in the b- and c-directions (Figure 12). These polarized spectra are shown in Figure 11. At room temperature, the absorption in the c-direction was much higher than for that with b-polarization. At 25,000 cm^{-1}, for example, the b-polarization is of the order of that in solution, whereas in the c-direction it has been enhanced ten-fold. In its intensity features, these spectra strongly resemble those of MGS, although there has not been a shift of bands to longer wavelengths from the solution spectra. At lower temperatures, the expected narrowing of bands has been observed, and for the c-polarization three components are easily resolvable with lower intensities than at room temperature. However, for the b-polarization—i.e., normal to the stacking direction—with a narrowing of the absorption band, there has been an increase in the height of the peak. This behavior is that of a symmetry-allowed transition of the sort which were observed in the low-temperature spectra of Mason and Gray (21). Since the molecule does not possess a center of symmetry, the possible effects of the asymmetric ligand field were examined. If a planar configuration was assumed, which is not strictly true because of puckering of the chelate ring, but which is not a bad approximation, an asymmetric field in the direction of the molecular dipole would provide some dipole-allowed intensity for the $d_{xz} \rightarrow d_{xy}^*$ transition. However, this would be for polarization in the z, or c-direction. Hence, the presence of an allowed transition normal to the chain direction would not be expected from this source.

Note that, as shown in Figure 12, the x and y axes have been directed between the ligands so that the antibonding or highest d-orbital is described as the d_{xy} rather than the $d_{x^2-y^2}$ which was used in the section on the PtCl$_4^{2-}$ ion. The assignment of the x and y axes is arbitrary between two equivalent choices in D_{4h} symmetry, and interchange of the choices interchanges the b_1 and b_2 representations. For this compound the present choice was convenient, since the x and y axes were then directed in the direction of the orthorhombic axes and the y axis was directed along the C_2 symmetry axis of the molecule.

The consequences of energy band formation from interactions of electrons in the platinum d-orbitals only were considered. For a set of the d-orbitals, say the d_{z^2} orbital on each platinum atom in the chain, were formed an LCAO wave function of the form

$$\psi_n(z^2) = \sum_{j=1}^{N} C_{n,j}(d_{z^2})_j \tag{7}$$

where

$$\sum_{j=1}^{N} C^2_{n,j} = 1 \qquad (8)$$

and $(d_{z^2})_j$ is the d_{z^2} orbital of the j atom in a chain of n-platinum atoms. From application of the simple theory of a particle in a one-dimensional box

$$C_{n,j} = \sqrt{\frac{2}{N+1}} \sin \frac{n \pi j}{(N+1)} \qquad (9)$$

where n can be an integer, $0 < n < N + 1$. With very large N, where only interactions between adjacent atoms are considered,

$$E_n = \int \psi_n H \psi_n d\tau = H_{jj} + 2H_{j,j+1}\cos \frac{n\pi}{N+1} \qquad (10)$$

Thus, a band of energy states based on the d_{z^2} orbitals is predicted. A corresponding set of wave functions, describing similar bands, can be written for each of the other four d-orbitals. With the atomic d^8 configuration of Pt^{2+}, the four lowest bands are filled; the highest band, based on the antibonding orbital involving d_{xy}, is empty. The expression for the transition moment involving a transfer of an electron from ψ_n in a filled band to ψ_m in the empty band requires an evaluation of the dipole moment integral

$$\rho_{n \to m} = \int \psi_n(z^2) r \, \psi_m(xy) \, d\tau \qquad (11)$$

From the symmetry properties of the products in these integrals, it can be concluded that transitions from the d_{z^2} band to the d^*_{xy} and from the $d_{x^2-y^2}$ band to the d^*_{xy} band are dipole-forbidden. However, a transition from the d_{xz} band to the d^*_{xy} band is dipole-allowed in the y polarization and from the d_{yz} band to the d^*_{xy} band is dipole-allowed in the x direction. Therefore, the band theory predicts a dipole-allowed transition normal to the chain from the bands arising from the orbitals which are degenerate (e_g) in D_{4h} symmetry. Since presumably interactions of the electrons between the platinum atoms are not large, the band is narrow and of low intensity. However, the band theory does account very nicely for the observed dipole-allowed transition in y polarization.

Direction of Current Research

In the preceding sections, a number of specific instances have been reviewed in which the absorption spectra of platinum group complexes have provided information concerning the chemical behavior of the systems and the electronic structure and bonding in the compounds. Some developments, published very recently, also deserve mention, since they illustrate the direction in which the research is moving.

Recently, Professor Schatz and coworkers at the University of Virginia have reported magnetic circular dichroism spectra for the $IrCl_6^{3-}$ ion in both solution and single crystals (*10, 22*). In addition, they have refined the theory, including the description of the *d*-orbitals, sufficiently that they can predict the sign of the MCD term. For these transitions the MCD arises from the *B*-term as defined by Stephens (*26*), and transitions of electrons from the t_{1u} and the t_{2u} orbitals (Figure 2) are predicted to have different signs. Their experiments therefore permit identification of the excited states for two of the charge transfer bands which are interchanged from the original suggestion by Jorgensen (*14*).

Dorain, Jordan, and their students at Brandeis University have recently reported some exciting crystal spectra. They dissolved small amounts of the hexahalo complexes as, for example, $OsCl_6^{2-}$, in host crystals such as the $PtCl_4$ (IV), the $ZnCl_4$ (IV), and the $HfCl_4$ (IV) salts of alkali metal ions (*7*). These crystals were cooled to liquid helium temperature and their spectra were recorded with high-resolution equipment. Now, instead of broad bands, they observed tremendous detail in their spectra. In the region from 17,000–35,000 cm^{-1}, well in excess of 100 lines were recorded. Portions of their spectra in very narrow regions are shown in Figure 13. The portion of the spectrum in the vicinity of 17,000 cm^{-1} corresponds to the weak band shown in Figure 1. The zero–zero line is identified (B) and is apparently excited by quadrupole excitation. Much stronger lines arise from the vibronic excitation of the lines indicated ν_4 and ν_3. The ν_4 and ν_3 represent the two t_{1u} normal vibrations of the octahedral ion. For many of the bands, the vibrational

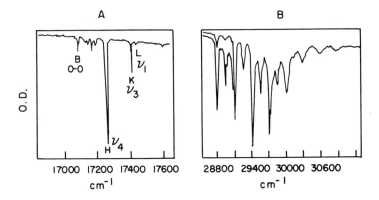

Figure 13. Spectra of the d^4 ion, $OsCl_6^{2-}$ dissolved in host crystals
A. Host is K_2PtCl_6. B. Host is Cs_2ZrCl_6. The O.D. scales are not equal in part A and B. Initial lines are enlarged by a factor of 5 in part B.

progressions permit an identification of the symmetry of the excited state. This quality of spectrum also affords an incentive for the theorist to tackle the difficult problem of the spectral intensities. In the vicinity of 29,000–30,000 cm^{-1} illustrated by the B portion in Figure 13, the molar absorptivity in solution is approximately 8000 cm^{-1} M^{-1}. However, from a consideration of intensities, the ligand field parameters, which include the strong spin-orbit coupling and the vibrational frequencies, Jordan et al. (13) conclude that all the bands in the region up to at least 35,000 cm^{-1} are ligand field—i.e., d-d transitions. This, of course, challenges the conclusion mentioned earlier that these bands arose from charge transfer $\pi \rightarrow t_{2g}$. Hence, some of the systematics of the band assignments need reevaluation.

Summary

The past two decades have witnessed a remarkable development in the experimental and theoretical considerations of the electronic absorption spectra of coordination complexes, of which the platinum metal compounds comprise an important segment. Despite their limited number of features, these spectra have proved useful for analytical purposes. With the development of ligand field theory and molecular orbital treatment, they have provided rather considerable information about the energy states of complexes, the energies which can be associated with particular orbitals, and the trends that occur in these energies between various elements, with various ligands and environments. The development of these theories has in turn stimulated experimental investigations, so that polarization of absorption bands, the magnetic circular dichroism, low temperatures, and the utilization of crystalline environments have greatly extended possibilities for identification of the transitions. Most certainly, the next few years will see a further application of these techniques, a refinement of the methods and instrumentation, and improved theoretical means for dealing with the problem of the energy states and the intensities.

Literature Cited

(1) Anex, B. G., Ross, M. E., Hedgecock, M. W., *J. Chem. Phys.* **1967**, 46, 1090.
(2) Basch, H., Gray, H. B., *Inorg. Chem.* **1967**, 6, 365.
(3) Chatt, J., Gamlen, G. A., Orgel, L. E., *J. Chem. Soc. (London)* **1958**, 486.
(4) Cotton, F. A., Harris, C. B., *Inorg. Chem.* **1967**, 6, 369.
(5) Day, P., Orchard, A. F., Thompson, A. J., Williams, R. J. P., *J. Chem. Phys.* **1965**, 42, 1973.
(6) Dickinson, R. G., *J. Am. Chem. Soc.* **1922**, 44, 2402.
(7) Dorain, P. B., Patterson, H. H., Jordan, P. C., *J. Chem. Phys.* **1958**, 49, 3845.

(8) Griffiths, J. H. E., Owen, J., Ward, I. M., *Proc. Roy. Soc. (London)* **1953**, A219, 526.
(9) Harris, C. M., Livingstone, S. E., Stephenson, N. C., *J. Chem. Soc. (London)* **1958**, 3697.
(10) Henning, G. N., McCaffery, A. J., Schatz, P. N., Stephens, P. J., *J. Chem. Phys.* **1968**, 48, 5656.
(11) Jacobson, Robert A., Benson, James E., private communication.
(12) Johannesen, R. B., Candela, G. A., *Inorg. Chem.* **1963**, 2, 67.
(13) Jordan, P. C., Patterson, H. H., Dorain, P. B., *J. Chem. Phys.* **1968**, 49, 3858.
(14) Jorgensen, C. K., *Acta Chem. Scand.* **1956**, 10, 500.
(15) Jorgensen, C. K., *Ibid.*, **1956**, 10, 518.
(16) Jorgensen, C. K., *Ibid.*, **1963**, 17, 1043.
(17) Jorgensen, C. K., *Mol. Phys.* **1959**, 2, 309.
(18) Jorgensen, C. K., "Absorption Spectra and Chemical Bonding in Complexes," Ch. 9, Pergamon, London, 1961.
(19) Ref. *17*, p. 114.
(20) Martin, D. S., Jr., Tucker, M. A., Kassman, A. J., *Inorg. Chem.* **1965**, 4, 1682.
(21) Mason, W. R., Gray, H. B., *J. Am. Chem. Soc.* **1968**, 90, 5721.
(22) McCaffery, A. J., Schatz, P. N., Lester, T. E., *J. Chem. Phys.* **1969**, 50, 379.
(23) McCaffery, A. J., Schatz, P. N., Stephens, P. J., *J. Am. Chem. Soc.* **1968**, 90, 5730.
(24) Miller, J. R., *J. Chem. Soc. (London)* **1965**, 713.
(25) Owen, J., Stevens, K. W. H., *Nature* **1953**, 171, 836.
(26) Stephens, P. J., *Inorg. Chem.* **1966**, 5, 491.
(27) Teggins, J. E., Gano, D. R., Tucker, M. A., Martin, D. S., Jr., *Ibid.*, **1967**, 6, 69.

RECEIVED December 16, 1969. Work performed in the Ames Laboratory of the U. S. Atomic Energy Commission, Contribution No. 2648.

8

NMR Spectra of Compounds of the Platinum Group Metals

R. STUART TOBIAS

Purdue University, Lafayette, Ind. 47907

> *Although resonances of the nuclei of platinum group metals have been observed in some cases, the small magnetic moments of many of these lead to very low NMR intensities, and the large spins of many of the nuclei cause very broad signals. Of much more use to the chemist have been studies on ^1H, ^{19}F, and ^{31}P resonances of ligands coordinated to the metals. Very high field resonances characterize hydride ligands. Since many compounds of these elements are stereochemically rigid, nuclear magnetic resonance spectroscopy is helpful in the elucidation of their structures. A number of the organometallic derivatives are nonrigid, and NMR has been used to study their reactions.*

The application of nuclear magnetic resonance to the study of the compounds of the platinum group metals has been limited by the properties of these nuclei. Data are collected in Table I for all of the isotopes of these elements which have nonzero spins.

The only nuclei which have spins of 1/2 and which consequently give relatively high resolution spectra are ^{103}Rh, ^{187}Os, and ^{195}Pt. All of the other isotopes have spins of 3/2 or greater and give very broad resonances because of the nuclear quadrupole moment. These finite quadrupole moments lead to rapid relaxation of the nuclei in environments not described by one of the cubic groups—i.e., when the coordination sphere has symmetry lower than T_d or O_h. As a result, spin–spin coupling usually is not observed with these nuclei.

Of the three isotopes with spins of 1/2, ^{187}Os occurs with very low natural abundance, 1.64%. This, combined with the very poor sensitivity of the nucleus compared to the proton, 7.93×10^{-5} at constant field, virtually precludes its study with contemporary spectrometers. Of all of these nuclei, only ^{103}Rh and ^{195}Pt lend themselves to NMR experiments.

Table I. Nuclear Properties of the Platinum Group Metals[a]

Isotope	NMR Frequency, 10 Kgauss Field, MHz	Natural Abundance, %	Sensitivity (Constant Field) Relative to $^1H = 1$	Spin, $h/2\pi$	Quadrupole Moment, $e \cdot 10^{-24} Cm^2$
^{99}Ru	1.9	12.81	1.07×10^{-3}	5/2	?
^{101}Ru	2.1	16.98	1.41×10^{-3}	5/2	?
^{103}Rh	1.340	100.	3.12×10^{-5}	1/2	–
^{105}Pd	1.74	22.23	7.94×10^{-4}	5/2	?
^{187}Os	1.8	1.64	7.93×10^{-5}	1/2	–
^{189}Os	3.307	16.1	2.34×10^{-3}	3/2	2.0
^{191}Ir	0.813	38.5	3.5×10^{-5}	3/2	1.5
^{193}Ir	0.86	61.5	4.2×10^{-5}	3/2	1.5
^{195}Pt	9.153	33.7	9.94×10^{-3}	1/2	–

[a] Reprinted from Ref. 57 by permission of Varian Associates.

Of the two, ^{195}Pt is somewhat easier to study because of its relatively good sensitivity and high resonance frequency.

Even with these limitations, nuclear magnetic resonance has made significant contributions to four areas of the chemistry of the platinum group metals: bonding problems, molecular stereochemistry, solvation and solvent effects, and dynamic systems—reaction rates. Selected examples in each of these areas are discussed in turn. Because of space limitations, this review is not meant to be comprehensive.

Bonding Problems

Conceptually, the simplest NMR experiment involves the study of the resonance of the metal nucleus itself, yielding chemical shift values and in some cases spin–spin coupling constant data. Recently, the internuclear double resonance technique (17) also has been used to measure these chemical shifts.

Figure 1 shows chemical shift data for a series of square planar platinum(II) complexes obtained by the INDOR technique (36). The proton resonance spectra of the ligands were monitored while sweeping through the ^{195}Pt resonant frequency range. The large range of chemical shifts, almost 1700 ppm from $(Me_2S)_2PtCl_2$ (low field) to $[(MeO)_3P]_4Pt^{2+}$ (high field) suggests that the shifts are dominated by the paramagnetic contribution. Qualitatively, this is supported by the observation that the complexes having low field resonances absorb in the blue region of the spectrum, while those with the highest field resonances are colorless.

Pidcock, Richards, and Venanzi (47) have conducted more detailed investigations of ^{195}Pt chemical shifts by direct measurement of the plati-

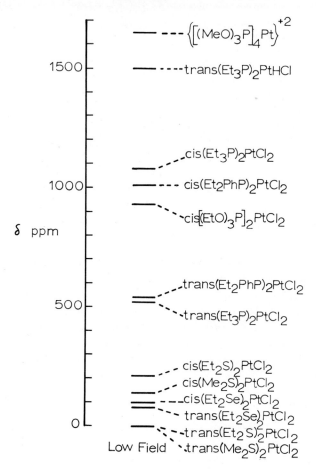

Figure 1. Chemical shifts in platinum(II) complexes; redrawn from Ref. 36

num resonances. The compound *trans*-[PtCl$_2$pepy$_2$], pepy = 4-pentylpyridine, was used as the reference. The general Expression 1 for nuclear shielding, σ, given by Ramsey (*50*) was simplified using a procedure first employed by Griffith and Orgel (*22*) to describe the shifts of cobalt(III)

$$\sigma = \sigma_d - 2\beta^2 \sum_{n \neq 0}(E_n - E_o)^{-1}[<\psi_o|\sum_i l_{iz}|\psi_n><\psi_n|\sum_k l_{kz}r_k^{-3}|\psi_o> + <\psi_o|\sum_k l_{kz}r_k^{-3}|\psi_n><\psi_n|\sum_i l_{iz}|\psi_o>] \quad (1)$$

complexes. In Equation 1, ψ_o and ψ_n are the wave functions of the ground and excited states, E_o and E_n are the corresponding energies, l_{iz} is the z component of the angular momentum operator for the ith electron, r_i is the distance of the ith electron from the nucleus, and β is the Bohr magneton. With Pt(II) complexes, the ground state is a singlet. If it is assumed that all of the complexes can be described in terms of D_{4h} symmetry, only singlet A_{2g} and E_g excited states need be considered, since these are the symmetry species of the angular momentum components under the operations of the D_{4h} point group. This leads to the simplified Expression 2 for the resonant frequency, ν. Here, H is the magnetic field strength, A is the diamagnetic shielding contribution, and B contains

$$\nu = -AH + 2BH\lambda\,^1A_{1g} \to\,^1A_{2g} + BH\lambda\,^1A_{1g} \to\,^1E_g \qquad (2)$$

$C_M{}^2(\overline{r^{-3}})$ as well as physical constants. The term C_M is the molecular orbital mixing coefficient for the platinum $5d$ orbital, and the r^{-3} average is taken over the platinum d radial function. The differences in energy of the ground and first excited A_{2g} and E_g states are given in terms of the transition wavelengths, λ.

If this simple model is adequate to represent the shielding in these platinum(II) complexes, a plot of $(\nu - \nu_{\text{ref}})$ vs. $(2\lambda\,^1A_{1g} \to\,^1A_{2g} + \lambda\,^1A_{1g} \to\,^1E_g)$ should be linear. Although a very rough linear relation is observed for complexes of the type PtX_2L_2, $X = $ halide, when X is constant and L is varied, different families of curves were obtained as X was varied from Cl⁻ to Br⁻ to I⁻. It seems likely that more electronegative ligands—e.g., Cl⁻—lead to a contraction of the platinum $5d$ orbitals and an increase in the B parameter through its r^{-3} dependence.

Dean and Green (14) also have obtained ¹⁹⁵Pt chemical shifts from double resonance studies of the square planar hydrides [Pt(PEt₃)₂HL], L = NO_3^-, NO_2^-, Cl⁻, Br⁻, I⁻, SCN⁻, CN⁻, and benzoate. They found little correlation between the optical properties—i.e., the spectrochemical series—and the sequence of chemical shifts and concluded that the excitation energies did not dominate the paramagnetic screening term. The shifts increased in the order NO_3^- < NO_2^- < Cl⁻ < −SCN⁻ < Br⁻ < CN⁻ < I⁻. This order is similar to the nephelauxetic series, and Dean and Green suggested that changes in the molecular orbital mixing coefficients were mainly responsible for the shifts.

In practice, it is not possible to separate the effect of changes in the MO mixing coefficients from those of orbital expansion and contraction, since they contribute to the same $C_M{}^2(r^{-3})$ term. Pidcock, Richards, and Venanzi chose to attribute the discrepancies to orbital contraction while Dean and Green considered only changes in the mixing coefficient. It seems certain that the large chemical shifts observed for these divalent platinum complexes arise mainly from the paramagnetic terms, but the simplified models used have been inadequate for a quantitative description of the shielding.

Considerably more information on the nature of the bonding in these complexes has been obtained from measurements of the spin–spin coupling of ^{103}Rh and more frequently ^{195}Pt to ^{1}H and ^{31}P nuclei in ligands. A useful review (40) of spin–spin coupling between directly bonded atoms including these has appeared recently. Phosphorus-31 spectra have yielded a wealth of data on cis (II) and trans (I) isomers of platinum(II) complexes where X = halide (46, 48). Grim, Keiter, and McFarlane (24) have obtained similar data for isomers of the complexes $(R_n(C_6H_5)_{3-n}P)_2PtCl_2$, R = CH_3, C_2H_5, C_3H_7, and C_4H_9; Grim and

$$\begin{array}{cc}
\text{P(C}_4\text{H}_9\text{)}_3 & \text{P(C}_4\text{H}_9\text{)}_3 \\
| & | \\
\text{X—Pt—X} & \text{(C}_4\text{H}_9\text{)}_3\text{P—Pt—X} \\
| & | \\
\text{P(C}_4\text{H}_9\text{)}_3 & \text{X} \\
\text{I} & \text{II}
\end{array}$$

Ference (23) have reported coupling constants for cis and trans isomers of Rh(III) complexes.

Data for a number of platinum(II) phosphine complexes are tabulated in Table II. The coupling constant for the cis isomer is ca. 1.5 times that for the trans isomer.

Table II. Platinum-195–Phosphorus-31 Spin–Spin Coupling Constants in Platinum(II) Complexes

$J(^{195}Pt-^{31}P)$, Hz.

Compound[a]	cis Isomer	trans Isomer	J_{cis}/J_{trans}	Ref.
$PtCl_2(Bu_3P)_2$	3508 ± 6	2380 ± 4	1.47	(46)
$PtBr_2(Bu_3P)_2$	3479 ± 10	2334 ± 8	1.49	(46)
$PtI_2(Bu_3P)_2$	3372 ± 4	2200 ± 4	1.53	(46)
$PtCl_2(Et_3P)_2$	3520	2400	1.47	(24)
$PtCl_2(Pr_3P)_2$	3530	2385	1.48	(24)
$PtCl_2(Bu_3P)_2$	3500	2392	1.46	(24)
$PtCl_2(Et_2PhP)_2$	3530	2482	1.42	(24)
$PtCl_2(Pr_2PhP)_2$	3561	2463	1.45	(24)
$PtCl_2(Bu_2PhP)_2$	3551	2462	1.44	(24)
$PtCl_2(BuPh_2P)_2$	3641	2531	1.44	(24)

[a] Bu = n-C_4H_9, Pr = n-C_3H_7, Et = C_2H_5, Ph = C_6H_5.

Consequently, a determination of J is sufficient for a structure assignment. In their original note, Pidcock et al. suggested that $J_{cis} > J_{trans}$ was caused by changes in σ bonding resulting from the synergic effect of π back bonding between platinum and phosphorus. In 1966, they used the Pople-Santry (49) Equation, 3, for the Fermi contact term in the expression for spin–spin coupling and concluded that the difference was caused by secondary σ orbital rehybridization effects rather than as a consequence of π bonding effects.

$$J \propto \gamma_M \gamma_X \Delta E^{-1} \alpha_M{}^2 \alpha_X{}^2 |\psi_M(0)|^2 |\psi_X(0)|^2 \quad (3)$$

The Fermi contact term seems to dominate the direct coupling between the heavy metals and ligand donor atoms. In Equation 3, the γ's are the nuclear magnetogyric ratios, ΔE is an average singlet–triplet excitation energy, α_M and α_X are the mixing coefficients for the metal valence s orbital and ligand valence s orbital in the hybrid orbitals used to describe the M–X bond, and $\psi_M(0)^2$ and $\psi_X(0)^2$ are the electron densities at the M and X nuclei, respectively. For coupling between ^{31}P and ^{195}Pt in the cis and trans isomers, the γ's are unchanged; and to a first approximation, it may be assumed that ΔE is constant. Orbital contraction effects are neglected; i.e., $\psi_M(0)$ and $\psi_X(0)$ are assumed to be the same in the two isomers. The change from cis to trans isomer should not have any significant effect on the phosphorus orbital for the bonding pair. Consequently, to a first approximation, $J \propto \alpha_M{}^2$; i.e., the coupling constant is proportional to the s character of the platinum hybrid used to bind the phosphine.

The spin–spin coupling data indicate that the platinum s character tends to be concentrated in the bonds to the more polarizable phosphine ligands. Pidcock et al. (46) adopted the term "trans influence" to describe the trans bond weakening effect observed with ligands such as phosphines. In the cis complex, the metal d and s character can be maximized in the bonds to the two phosphines, leaving the trans bonds with predominantly $6p$ character consistent with their ease of substitution and long bond lengths (56).

Similar measurements of the ^{195}Pt–^{31}P coupling in phosphine complexes of platinum alkyls (1) again indicate that the s character tends to concentrate in the bonds to the most polarizable ligands, the alkyl groups. Structures III (24), IV (1), and V (1) show the effect of increasing methylation of PtCl$_2$[P(C$_2$H$_5$)$_3$]$_2$. The bond to a ligand trans to a methyl group is weakened, consistent with a concentration of the platinum $6p$ character in these bonds. Similar effects have been reported for aryl, silyl, and germyl derivatives of platinum(II) (25). In general, the J ^{195}Pt–^{31}P values are small when phosphorus is trans to a ligand of high trans influence—e.g., a phosphine or a carbanion—and large when trans to a relatively nonpolarizable ligand like chloride.

Similar effects are observed with octahedral complexes where $J_{cis} \approx 1.5\, J_{trans}$ as illustrated in structures VI and VII (46). The ratio of the

$$\underset{\text{VI}}{J = 1462 \pm 4 \text{ Hz.}} \qquad \underset{\text{VII}}{J = 2070 \pm 2 \text{ Hz.}}$$

coupling constants for Pt(IV):Pt(II) = 0.59 for the cis complexes II and VII and 0.61 for the trans complexes I and VI. The fact that these values are near that of the ratio of s character for d^2sp^3 and dsp^2 hybrids, $1/6 : 1/4 = 0.67$, suggests that $|\psi_{Pt}(0)|^2$ is comparable for both oxidation states.

The similarity of the behavior of Pt(II) and Pt(IV) complexes was used by Pidcock et al. as evidence that only σ bonding effects were responsible for the cis–trans difference. It is generally assumed that d_π–d_π back bonding is negligible with Pt(IV), but this is based on Pt(IV) d orbitals being contracted with respect to Pt(II) orbitals. The evidence also cited for the invariance of $|\psi_{Pt}(0)|^2$ with oxidation state suggests that d_π–d_π back bonding should be comparable in Pt(II) and Pt(IV) complexes.

Parshall (44, 45) has used fluorine-19 nuclear magnetic resonance spectra of the fluorophenyl derivatives of platinum(II), VIII and IX, to study the trans influence of the X ligand. The meta ^{19}F shielding

VIII   IX

parameter varies with the σ donor character of the fluorophenyl ring substituent, while the para shielding parameter is affected markedly by the π donor character of the substituent. Consequently, the para shielding parameter reflects the ability of the ligand X bound to platinum to compete with the p-fluorophenyl group for platinum d_π electron density. To correct for simple inductive effects, the difference between the para and the meta shielding parameters was considered. In this way, the ligands X were classified in increasing order of π acceptor character as $Cl^- \simeq I^- < Br^- < OCN^-$ or $NCO^- \simeq CH_3 < -SCN^-$ or $-NCS^- < C_6H_5 \sim p\text{-}FC_6H_4 \simeq m\text{-}FC_6H_4 < SnCl_3^- \simeq C_6H_5-C\equiv C < CN^-$.

There have been some determinations of the signs of the spin–spin coupling constants for complexes of platinum. As might be expected, the coupling constants $^2J(^{195}Pt-CH_2)$ and $^3J(^{195}Pt-CH_3)$ have opposite signs in $[(C_2H_5)_3PtCl]_4$ (32), and the absolute values are 86 and 72.0 Hz, respectively. Heteronuclear double resonance measurements on cis and trans dichlorobis(triethylphosphine)platinum(II) (39) indicate that $^1J(^{195}Pt-^{31}P)$ is positive, $^2J(^{31}P-CH_2)$ is negative, and $^3J(^{195}Pt-CH_2)$ is positive. Similarly, heteronuclear double resonance spectra of PtHCl(PEt$_3$)$_2$ (35) show that $^1J(^{195}Pt-^{31}P)$ and $^1J(^{195}Pt-^1H)$ are of the same sign, while $^2J(^{31}P-^1H)$ is of opposite sign.

Stereochemistry

The principal application of NMR with compounds of the platinum group metals has been in the study of their stereochemistry. Cis and trans isomers of the Pt(PR$_3$)$_2$L$_2$ and Pt(PR$_3$)$_2$X$_4$ type are identified easily because of the difference in the coupling constants as discussed above.

The tris(tri-n-butyl)phosphine complex of rhodium trichloride has the trans (mer) structure X (23). Two ^{31}P resonances are observed with

$$(C_4H_9)_3P_b \diagdown \underset{\underset{P_a(C_4H_9)_3}{|}}{\overset{\overset{Cl}{|}}{Rh}} \diagup P_b(C_4H_9)_3 \diagdown Cl$$

X

an intensity ratio of 2:1. Both signals are split into doublets by coupling with ^{103}Rh, $J(^{103}Rh-^{31}P_a) = 114$, $J(^{103}Rh-^{31}P_b) = 84$ Hz. Again the ratio of coupling constants for phosphorus trans to chlorine relative to phosphorus trans to phosphine is ca. 1.5 This spectrum also exhibits small splittings of the more intense phosphorus resonance by 21 Hz caused by coupling through two bonds with the unique phosphorus-31 nucleus P_a. The splitting of the less intense resonance by the two equivalent phosphorus nuclei was not resolved.

Phosphorus-31 proton spin–spin coupling constants determined with proton resonance spectra have been used to distinguish cis and trans isomers of hydrido complexes containing phosphine ligands. For example, the complexes $IrH_xY_{3-x}L_3$ (Y is a uninegative anion, L is a neutral phosphine ligand) have $^2J(^{31}P-H) = $ 11–21 Hz for cis and 130–163 Hz for trans H and L ligands (6). Analogous data have been used to assign structure XI to the product of the reaction between $RhCl(PPh_3)_3$ and HCl in solution (2). The hydride resonance gives a pair of overlapping triplets centered at $\tau = 26.1$ ppm because of coupling with the rhodium

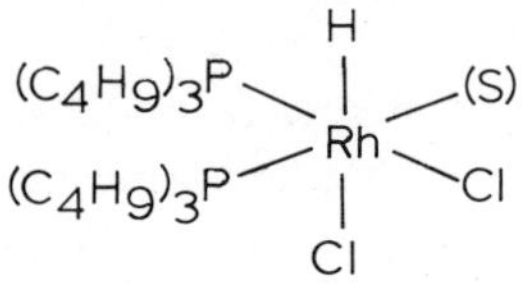

```
            H
(C4H9)3P\   |   /(S)
         \  |  /
          Rh
         /  |  \
(C4H9)3P/   |   \Cl
            Cl
```

XI

nucleus and two equivalent phosphorus nuclei. The assignment of the hydride cis to the phosphines is based on the small value of $^2J(^{31}P-^1H) = 17$ Hz.

Morgan, Rennick, and Soong (41) used coupling between platinum-195 and the hydroxylic protons of $[(CH_3)_3PtOH]_4$, XII, to show that the

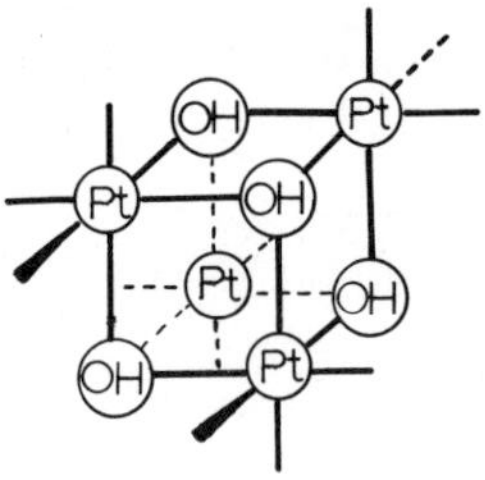

XII

hydroxide was isostructural with the chloride, bromide, and iodide. Since platinum is 33.7% ^{195}Pt, while all other naturally-occurring isotopes have zero spin, a given hydroxo group bridges between 3 platinums where 1, 2, or all 3 may have spins of 1/2. The predicted proton resonance spectrum for cage structure XII is illustrated in Figure 2, and this agrees with the observed spectrum when $^2J(^{195}Pt-OH) = 11.3$ Hz, $\tau = $ ca. 11.5 ppm.

The magnitude of coupling between the methyl protons and platinum-195 also is helpful in assigning structures to trimethylplatinum(IV) complexes. The $^2J(^{195}Pt-CH_3)$ coupling constant appears to be influenced primarily by the nature of the ligand trans to the methyl group, and the solvent used makes little difference (33, 54). This same effect

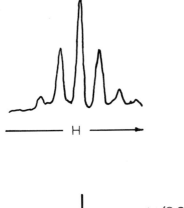

		$1 \times (0.662)^3$
	| |	$3 \times (0.662^2 \times 0.337)$
ı	| ı	$3 \times (0.662 \times 0.337^2)$
.	. . .	$1 \times (0.337)^3$

Inorganic Chemistry

Figure 2. Predicted spectrum for the hydroxylic proton resonance of $[(CH_3)_3PtOH]_4$; observed spectrum redrawn from Ref. 41

has been observed for spectra of the complex cations $[(CH_3)_3Pt(OH_2)_n\text{-}py_{3-n}]^+$ (py = pyridine) determined with aqueous solutions (7). Only two coupling constants were observed for mixtures of these complexes with values of 79.7 Hz for methyls trans to water molecules and 67.7 Hz for methyls trans to pyridine. Structures XIII to XIX illustrate average values for the $^2J(^{195}Pt–CH_3)$ coupling constants as the ligand trans to the methyl groups are varied (33).

XIII: trans-Cl, $J^{195}Pt\text{-}^1H = 81.7$
XIV: trans-Br, 80.1
XV: trans-O, 79.1 Hz
XVI: trans-I, $J^{195}Pt\text{-}^1H = 78.0$
XVII: trans-C, 75.9
XVIII: trans-N, 74.9 Hz

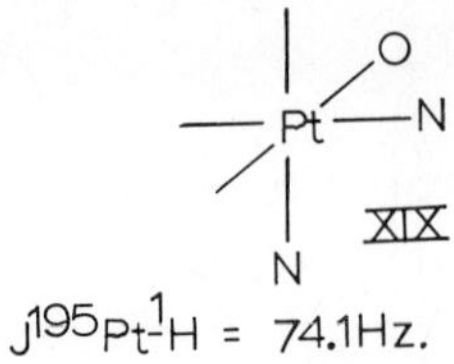

$$\begin{array}{c} \mathrm{-Pt-N} \\ | \\ \mathrm{N} \end{array} \quad \text{XIX}$$

$J^{195}\mathrm{Pt}\text{-}^1\mathrm{H} = 74.1\,\mathrm{Hz}.$

One application of these data is illustrated in XX. Trimethylplatinum(IV) oxinate is dimeric. The molecule has a two-fold axis in solution, and three methyl proton resonances with coupling constants of 80.4, 76.0, and 70.3 Hz are observed. The signal with the smallest J is assigned

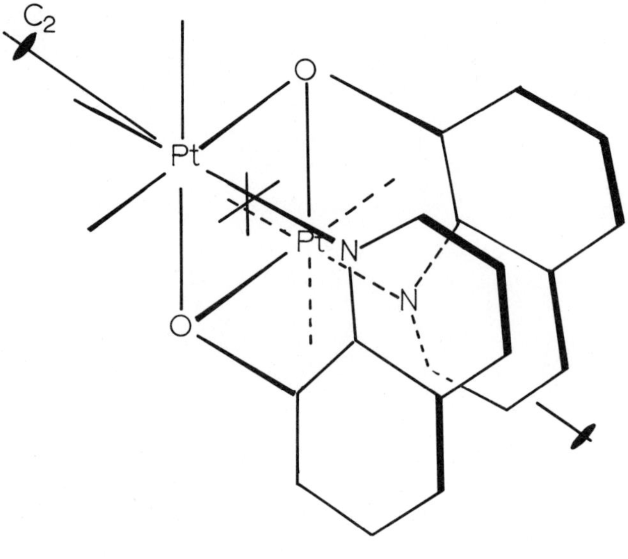

XX

to the methyl protons trans to nitrogen and the other two signals to methyls trans to oxygen. The spectrum shows that the bonds from Pt to the two bridging oxygens are not equivalent. The crystallographic values (34) were 2.22 ± 0.03 and 2.29 ± 0.04 A, and the errors in the light atom positions were sufficiently large to leave the nonequivalence of the bridge bonds in doubt.

Structures can be assigned to square planar complexes of the type $\mathrm{PdX_2(PR_3)_2}$ and $\mathrm{PtX_2(PR_3)_2}$ (X = uninegative anion) on the basis of proton rather than phosphorus-31 resonance by using dimethylphenylphosphine as the ligand (27, 29). The free ligand shows a doublet methyl proton resonance, $\tau = 8.61$ ppm, $^2J(^{31}\mathrm{P}\text{-}^1\mathrm{H}) = 1.7$ Hz, because of coupling with the phosphorus nucleus. In the cis complexes, a sharp doublet also is observed $^2J(^{31}\mathrm{P}\text{-}^1\mathrm{H}) = 7$ to 13 Hz. If the two dimethylphenylphosphines are mutually trans, the resonance is usually a well resolved triplet because of coupling of the methyl protons with both of the phosphorus nuclei. These complexes are best treated as $AA'X_3X_3'$ systems (39) with very small $^{31}P \ldots ^{31}P$ coupling constants in the cis isomers, while for the trans isomer $J(^{31}P \ldots ^{31}P) \gg 10$ Hz. Consequently, determination of the PMR spectrum suffices for the structure assignment. A similar coupling scheme has been observed for octahedral iridium(III) compounds of the type $[IrX_2Y(PMe_2Ph)_3]$ (X and Y are uninegative anions) (28).

Proton NMR has been used to study the geometrical isomers of the 2,5-dithiahexane complexes of $RhCl_3$ in D_2O solution. Cationic species $[RhCl_2L_2]^+$ are obtained, and the trans arrangement of chlorides was assigned on the basis of vibrational spectra (58). Because of the possible dithiahexane ligand conformations, up to five different isomers can exist with trans chlorides. The proton resonance spectrum shows a series of doublets with $^3J(^{103}Rh-^1H) = 1.05$ Hz. By irradiation at the proper ^{103}Rh resonance frequency, the doublets can be collapsed one by one, giving the ^{103}Rh chemical shifts in the different complexes (38). Seven different species were identified with ^{103}Rh chemical shifts over the range 0–296 ppm, indicating that some cis isomers also were present. Structure XXI was assigned to the isomer with a shift of 38 ppm relative to the lowest field resonance, because the spectrum showed three different methyl resonances with one twice the intensity of the other two, consistent with this geometry.

XXI

Solvation and Solvent Effects

Some studies using oxygen-17 NMR spectroscopy have been made with aqueous solutions of platinum(II) and organoplatinum(IV) cations. Many years ago, it was reported that treatment of $PtCl_2(NH_3)_2$ with an aqueous solution of $AgNO_3$ or Ag_2SO_4 gave concentrated solutions of a species presumed to be the diaquodiammineplatinum(II) cation (30). Solutions of the perchlorate salt show a bound water oxygen-17 resonance signal 92.6 ± 1 ppm upfield from the bulk water resonance (20). This large chemical shift for the water bound to platinum may be compared with the value of *ca.* −11 ppm for $Al(ClO_4)_3$ solutions (9). There was no change in the chemical shift over the temperature range 29° to 85°, indicating that the water is relatively inert to substitution. Integration of the bound and bulk water resonances gave the hydration number as 1.9 ± 0.1, confirming the formulation of the cation as $[(H_3N)_2Pt(OH_2)_2]^{2+}$. Undoubtedly, this cation is sufficiently inert that the solvation number could be determined by conventional—*e.g.*, isotope dilution—techniques.

The hydration of the trimethylplatinum cation also has been studied by ^{17}O NMR of aqueous solutions of $(CH_3)_3PtClO_4$ (20, 21). Although many platinum(IV) complexes are numbered among the most inert coordination compounds known, the water molecules bound to $(CH_3)_3Pt^+$ exchange rapidly with the bulk solvent at room temperature. Only near 0°C is it possible to record a bound water resonance. Employing the molal shift of the bulk water solvent caused by added $Dy(ClO_4)_3$, it was possible to determine the hydration number to be 3.0 ± 0.1. The cation is formulated correctly as octahedral $[(CH_3)_3Pt(OH_2)_3]^+$. Because of the large quadrupole moment of the ^{17}O nucleus, $I = 5/2$, no coupling with the ^{195}Pt nucleus was detected in any of the spectra described above. These experiments illustrate the very great lability of some of the organo derivatives of the platinum group metals.

Clegg and Hall (8) noted that the addition of sulfuric acid to an aqueous solution of $[(CH_3)_3Pt(NH_3)_3]Cl$ caused a shift of 0.5 ppm in the methyl proton resonance and an increase in the value of $^2J(^{195}Pt-CH_3)$ from 71.0 to 79.7 Hz because of rapid formation of the triaquo species. Addition of an excess of the ligands H_3CNH_2, pyridine, SCN^-, NO_2^-, and CN^- all gave rise to spectra characteristic of the complexed trimethylplatinum(IV) cation (7).

Sawyer and coworkers have studied the proton resonance spectra of aqueous solutions of the 1:1 and 1:2 complexes formed from Pt(II) (51), Pd(II) (53), and Rh(III) (52) with nitrilotriacetic acid, N-methyliminodiacetic acid, and iminodiacetic acid as a function both of temperature and of pH. The acetate methylene protons exhibit an AB pattern in all of the complexes, indicating that the lifetimes of metal–oxygen and metal–nitrogen bonds are relatively long—*i.e.*, that the complexes are stereochemically rigid. Both the mono XXII and bis XXIII complexes of platinum(II) show $^{195}Pt-^1H$ couplings through three bonds to the acetate methylene protons and to the N-substituted groups. No analogous couplings were found with the Rh(III) chelates. Complex changes occur as the pH is raised with the concomitant formation of hydroxo complexes and probably polynuclear species. These changes can be followed conveniently with the proton resonance spectra, although a detailed interpretation of the effects is very difficult.

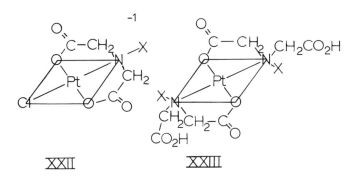

XXII   XXIII

The interaction of cations formed from the platinum group metals with solvents like acetonitrile, dimethylsulfoxide, and other molecules containing methyl protons is studied easily with ^{1}H NMR, since these complexes are rather inert. O'Brien, Glass, and Reynolds (42) determined the solvation number of $(H_3N)_2Pt^{2+}$ in acetonitrile solution from the proton resonance spectra of solutions of anhydrous $(H_3N)_2Pt(ClO_4)_2$. Both bound and bulk acetonitrile resonances were observed with no evidence for any exchange broadening to 81°, the boiling point of acetonitrile. Coupling of the ligand protons with the platinum nucleus leads to the usual 1:4:1 triplet, $^4J(^{195}Pt-^1H) = 12.1$ Hz. Integration of one component of the triplet for the bound solvent relative to the ^{13}C side band of the bulk solvent gave a solvation number of 2.0 ± 0.1. It also was possible to determine approximately the relative integrals of the amine and acetonitrile protons, although the amine resonances are very broad because of the ^{14}N quadrupole moment. This method gives a solvation number of 2.0 ± 0.1, too. There was no shift in the bulk solvent resonance as the concentration of $[(H_3N)_2Pt(NCCH_3)_2](ClO_4)_2$ was varied, suggesting that there are no significant solvent interactions in the axial positions of platinum.

Anhydrous $(H_3N)_3Pt(ClO_4)_2$ also has been studied in dimethylsulfoxide solution (55), since Me$_2$SO is an ambidentate ligand and probably coordinates via sulfur rather than oxygen as is known to be the case with palladium. The methyl proton resonance for the bound DMSO is shifted 2.54 ppm upfield from the bulk solvent, and the ligand exchange rate is negligible at 35°. Integrations of the bulk and bound DMSO signals or the bound DMSO and amine proton resonances both gave solvation numbers of 2.0 ± 0.1. The coupling constant $J(^{195}Pt-^1H) = 23.2$ Hz, and this large a value suggests that DMSO is bound via sulfur and not oxygen—i.e., that the coupling is through only three and not four bonds. Undoubtedly, isolation of the crystalline adducts $(H_3N)_2Pt(ClO_4)_2 \cdot 2B$ from solution would confirm these stoichiometries; however, these compounds are very explosive.

McFarlane and White (37) have noted that dimethylsulfoxide will replace Me$_2$S and Me$_2$Se when [(Me$_2$L)$_2$PtCl$_2$] (L = S, Se) complexes are dissolved in DMSO according to Reactions 4 and 5.

$$(Me_2L)_2PtCl_2 + DMSO \rightleftarrows Me_2L + (DMSO)(Me_2L)PtCl_2 \quad (4)$$

$$(DMSO)(Me_2L)PtCl_2 + DMSO \rightleftarrows Me_2L + (DMSO)_2PtCl_2 \quad (5)$$

Well-resolved ^{195}Pt–^{1}H coupling (DMSO protons) of 21 to 24 Hz is observed, indicating that the DMSO exchanges slowly with the bulk solvent.

Proton magnetic resonance spectra have been used to assign structures to the geometrical isomers of RhCl$_3$(NCCH$_3$)$_3$, since the cis complex gives a single resonance while the trans isomer gives two resonances separated by 0.05 ppm (5). The stoichiometry in acetonitrile solution of the complex [(C$_2$H$_5$)$_4$N][RhCl$_4$(NCCH$_3$)$_2$] was established by integration of the bound acetonitrile resonances with respect to the tetraethylammonium methyl or methylene signals.

Since most of the complexes of the platinum group metals are stereochemically rigid and ligands exchange very slowly, studies of stereochemistry with solutions are straightforward. Care must be used, however, in interpreting spectra of organic derivatives of these metals, since many of them are not rigid.

Dynamic Systems

The NMR method for studying the rates of moderately fast reactions has found little use with the strictly inorganic complexes of the platinum group metals, since they include many of the most inert complexes known. There are, however, two types of compounds of these metals which often are rather labile—i.e., the organo and hydrido derivatives. For such compounds, the NMR method, although less useful for stereochemical studies, is proving very valuable for studying reaction rates.

The trimethylplatinum(IV) species provide a particularly clear example of the effects described above. Octahedral inorganic complexes of platinum(IV) are very inert to ligand substitution, and the fact that such reactions occur at all is often attributable to catalysis by platinum(II) impurities (3). The organometallic complexes [(CH$_3$)$_3$PtL$_3$] undergo substitution reactions of the L ligands very rapidly as a result of the coordination of the three polarizable "methide" ligands. Glass and Tobias (20) studied the rate of water exchange between [(CH$_3$)$_3$Pt(OH$_2$)$_3$]$^+$ and the bulk solvent using oxygen-17 NMR and found the average lifetime of a water molecule bound to (CH$_3$)$_3$Pt$^+$ to be only *ca.* 9×10^{-5} sec at 25°. Approximate expressions were used to relate the line shapes to the pseudo first order rate constant.

The first applications of NMR to the study of dynamic systems of the platinum group metals appear to have been studies on the rotation about the metal–olefin bond of coordinated olefins, and this process has been investigated by many workers. There are two reasonable orientations of an olefin with respect to the rest of a square planar complex, XXIV and XXV.

XXIV   XXV

Most of the ground states of complexes seem to have structure XXIV, but XXV reasonably could provide a mechanism for rotation about the metal–olefin bond axis with a low energy barrier. Cramer (11) found that π-cyclopentadienylbis(ethylene)rhodium(I), XXVI, gave two broad signals (τ = 7.23, 8.88 ppm) for the ethylene protons at $-25°$ and that

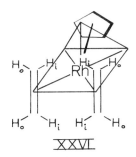

XXVI

upon heating these coalesced to a single resonance (τ = 8.07 ppm) at 57°. The most likely process which would lead to equivalence of all of the ethylene protons is rotation of the ethylene molecules about the rhodium–olefin bond, since the complex exhibits no measurable exchange with free ethylene after heating for five hours at 100°.

In contrast to the results with the cyclopentadienyl derivative, acetylacetonatobis(ethylene)rhodium(I), XXVII shows a single coalesced signal for the ethylene protons at 25°, and only upon cooling to $-58°$ are two signals observed at τ = 6.42 and 7.49 ppm (11). If ethylene is added to the cold solution, only a single resonance for bound and free ethylene is observed at τ = ca. 6.95 ppm. Thus, the ethylene exchanges rather rapidly even at $-58°$, and a bimolecular process is indicated. The anion of Zeise's salt, $[(C_2H_4)PtCl_3]^-$, and free ethylene in methanolic HCl also exhibit only one proton resonance at temperatures as low as $-75°$ (10),

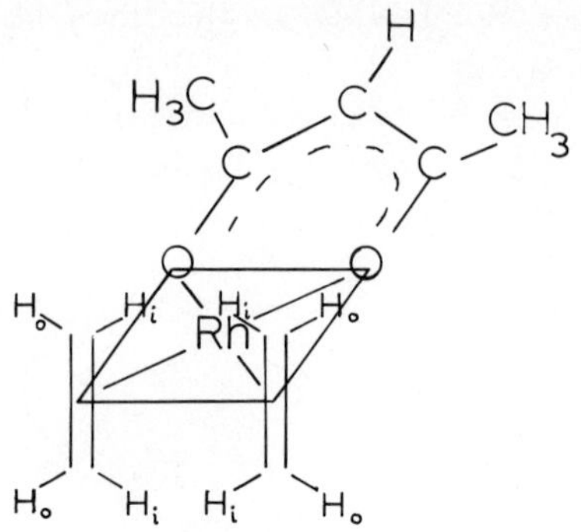

XXVII

and the τ value is a function of the $[(C_2H_4)PtCl_3]^-:C_2H_4$ mole ratio. Consequently, exchange is very rapid with this complex, too. One possible explanation of the fact that $(\pi\text{-}C_5H_5)Rh(C_2H_4)_2$ does not exchange ethylene at a measurable rate is that the π-cyclopentadienyl group functions as a six-electron donor, making the complex much less susceptible to nucleophilic attack (*11, 12*).

Reaction of $(\pi\text{-}C_5H_5)Rh(C_2H_4)_2$ with SO_2 yields $(\pi\text{-}C_5H_5)Rh(C_2H_4)SO_2$. Again, this compound shows only a single ethylene proton resonance at $-2°$, $\tau = 6.68$ ppm, but upon cooling to $-50°$, two resonances typical of an A_2B_2 spectrum are observed (*12*). Recently, Cramer, Kline, and Roberts (*13*) reinvestigated $(\pi\text{-}C_5H_5)Rh(C_2H_4)_2$, XXVI, and $(\pi\text{-}C_5H_5)Rh(C_2H_4)SO_2$ using more precise equations for the variation of line shape with lifetime and obtained 15.0 ± 0.2 and 12.2 ± 0.8 kcal/mole, respectively, for the activation energies for rotation. With $(\pi\text{-}C_5H_5)Rh(C_2F_4)(C_2H_4)$, the barrier was 13.6 ± 0.6 kcal/mole. The reduction of the barrier to rotation upon substitution of electron-withdrawing groups, *e.g.*, C_2F_4, is consistent with a reduction in back bonding.

Brause, Kaplan, and Orchin (*4*) examined the barrier to rotation of styrene and *tert*-butylethylene in 2,4,6-trimethylpyridine complexes of platinum(II), XXVIII.

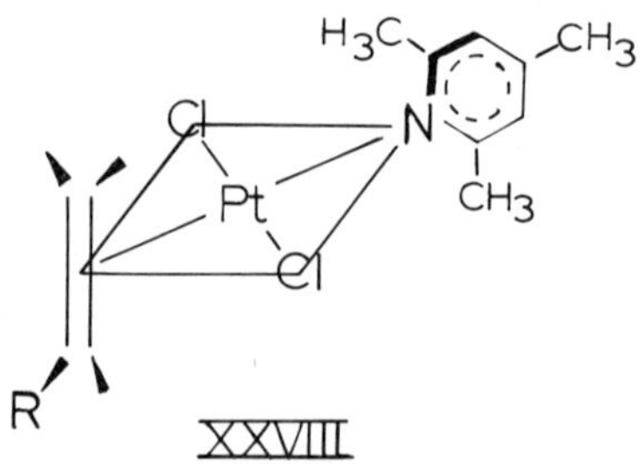

XXVIII

The proton resonance of the o-methyl protons of the 2,4,6-trimethylpyridine ligand bound trans to styrene is a single triplet [$J(^{195}Pt-CH_3)$ = 13 Hz] at room temperature, but it splits into two overlapping triplets at −60°. Similar behavior is found with *tert*-butylethylene. Since the coupling between platinum and the olefin protons ($J^{195}Pt-^1H$ = ca. 10 Hz) is maintained for the coalesced signal, the process was assumed to involve internal rotation about the ethylene–platinum bond; however, it now is known that the pyridine ligands are bound quite weakly because of the large trans influence of ethylene. Kaplin, Schmidt, and Orchin (*31*) studied the exchange of substituted pyridines and olefin in complexes analogous to XXVIII. At room temperature, the pyridine bases exchange rapidly on the PMR time scale, probably by both dissociative and bimolecular mechanisms. The olefins are still sufficiently strongly bound to give spin–spin coupling with platinum-195.

Holloway, Hulley, Johnson, and Lewis (*26*) have observed rotation of several olefins in the platinum(II) complexes of type XXIX. For the

XXIX

ethylene protons, two overlapping triplet resonances (^{195}Pt coupling) are observed at low temperatures, τ = 5.48 and 5.53 ppm. Coalescence occurs at −28°C. The ^{195}Pt coupling persists after coalescence. Free energies of activation for the rotational process range from 10.9 kcal/mole with tetramethylethylene to 15.8 kcal/mole with trans but-2-ene.

A number of studies have been made of the exchange of coordinated and free ligands. Eaton and Suart (*16*) used ^{31}P NMR to examine the extent of dissociation of chlorotris(triphenylphosphine)rhodium(I) in CH_2Cl_2 solution. At room temperature, the complex shows two ^{31}P resonances for the trans and the two cis phosphines, and these are split by interaction with the ^{103}Rh and ^{31}P nuclei. Addition of excess triphenylphosphine has no effect on the complex spectrum, and exchange is slow on the NMR time scale. Dissociation, 6, which was suggested on the basis

$$\text{Rh}(\text{PPh}_3)_3\text{Cl} + \text{CH}_2\text{Cl}_2 \rightleftarrows \text{Rh}(\text{PPh}_3)_2\text{Cl}(\text{CH}_2\text{Cl}_2) + \text{PPH}_3 \qquad (6)$$

of molecular weight data (*43*), occurs only to the extent of less than 5% at concentrations $> 10^{-2} M$. Mixtures of tris(*p*-tolyl)phosphine and tris-(triphenylphosphine)rhodium chloride give *p*-tolyl methyl resonances owing to tris(*p*-tolyl)phosphines bound both cis and trans at $-35°$. These are collapsed at 0°, and at 35° in the presence of excess tris(*p*-tolyl)phosphine, the ligand resonance is averaged over both bound and free phosphine. The activation energy for the intramolecular exchange of the cis and trans phosphines is *ca.* 6 kcal/mole, and this process is appreciably faster than the exchange between free and bound phosphine.

Deeming and Shaw (*43*) have studied the exchange of dimethylphenylphosphine with complexes of the type trans RhX(CO)(PR$_3$)$_2$, X = halide, XXX. The ligand methyl resonance gives a triplet because of coupling with both phosphorus nuclei, which is split further by coupling

XXX

with the rhodium nucleus $^3J(^{103}\text{Rh}-^1\text{H}) = 1$ Hz. See the discussion of coupling in trans bis(phosphine) complexes in the section on stereochemistry. Addition of sufficient dimethylphenylphosphine to give a 1:10 ligand:complex mole ratio causes the triplet to collapse because of exchange of bound and free phosphine. Only a broad singlet is observed. The vanishing of the coupling with ^{31}P, $I = 1/2$, indicates that $^2J(^{31}\text{P}-^1\text{H})_{\text{bound}}$ and $^2J(^{31}\text{P}-^1\text{H})_{\text{free}}$ are of opposite sign and that they average to zero. This is supported by the observation that the addition of more phosphine leads to the expected doublet arising from coupling with ^{31}P.

Fackler and coworkers (*18, 19*) have used the methyl proton resonances to determine activation energies for the exchange of methyldiphenylphosphine coordinated to the bis-complexes of platinum(II) with ligands such as *p*-dithiocumate (*18*), 3,4,5-trimethoxydithiobenzoate (*18*), XXXI, and *o*-benzylxanthate (*19*).

XXXI

The $J(^{195}\text{Pt}-^1\text{H})$ coupling, 39–40 Hz, shows that the phosphine is coordinated to platinum. Since it occupies an axial position of the basically square planar complex, the phosphine is rather labile. Activation energies for exchange of 4.2 and 19.7 kcal/mole were found where the ligand was *p*-dithiocumate and 3,4,5-trimethoxydithiobenzoate, respectively.

Yagupsky and Wilkinson (59) have examined the hydride resonance of the 5-coordinate hydridodicarbonylbis(triphenylphosphine)iridium(I) as a function of temperature. Two isomers appear to exist in equilibrium, since the coupling constant $^2J(^{31}\text{P}-^1\text{H})$ is considerably smaller than expected for coupling to a cis phosphine (1.5–8.5 Hz, depending upon solvent at 35°), and it is temperature-dependent. See the discussion of hydride–phosphine coupling in cis and trans complexes in the section on stereochemistry. Since the coupling decreases to zero as the temperature is lowered and then increases again as the temperature drops still further, it appears that the two isomers have $^{31}\text{P}-^1\text{H}$ coupling constants of opposite sign. As the equilibrium between the isomers is shifted by varying the temperature, there exists a temperature at which J averages to 0; i.e., $\chi_A J_A + \chi_B J_B = 0$ where χ_A and χ_B are the mole fractions of the two isomers and J_A and J_B are their $^{31}\text{P}-^1\text{H}$ coupling constants.

Some caution should be used in the interpretation of the activation parameters obtained in many of these studies. Particularly in the earlier work, approximate equations relating line shape to rate parameters often have been used, so many of the rates are basically order of magnitude values. The current use of computers to determine the best agreement between calculated and observed spectra should improve the accuracy of these calculations. Other factors such as changes in the chemical shifts between two resonances with temperature can cause large errors in the activation energy.

Summary

The principal application of NMR in the study of the compounds of the platinum group metals is likely to continue to be in structure determination. As spectrometers suitable for heteronuclear double resonance measurements become increasingly available, more information on the chemical shifts of the metal nuclei will be produced. NMR should prove extremely useful for the study of labile complexes of these elements, and it is precisely these systems which are of interest as catalytic intermediates. This technique affords a convenient method for the study of intra- and intermolecular exchange processes.

Literature Cited

(1) Allen, F. H., Pidcock, A., *J. Chem. Soc.* **1968**, A, 2700.
(2) Baird, M. C., Mague, J. T., Osborn, J. A., Wilkinson, G., *J. Chem. Soc.* **1967**, A, 1347.
(3) Basolo, F., Pearson, R. G., "Mechanisms of Inorganic Reactions," 2nd ed., p. 238, Wiley, New York, 1966.
(4) Brause, A. R., Kaplan, F., Orchin, M., *J. Am. Chem. Soc.* **1967**, 89, 2661.
(5) Catsikis, B. D., Good, M. L., *Inorg. Chem.* **1969**, 8, 1095.
(6) Chatt, J., Coffey, R. S., Shaw, B. L., *J. Chem. Soc.* **1965**, 7391.
(7) Clegg, D. E., Hall, J. R., *Australian J. Chem.* **1967**, 20, 2025.
(8) Clegg, D. E., Hall, J. R., *Spectrochim. Acta* **1967**, 23A, 263.
(9) Connick, R. E., Fiat, D. N., *J. Chem. Phys.* **1963**, 39, 1349.
(10) Cramer, R., *Inorg. Chem.* **1965**, 4, 445.
(11) Cramer, R., *J. Am. Chem. Soc.* **1964**, 86, 217.
(12) *Ibid.*, **1967**, 89, 5377.
(13) Cramer, R., Kline, J. B., Roberts, J. D., *J. Am. Chem. Soc.* **1969**, 91, 2519.
(14) Dean, R. R., Green, J. C., *J. Chem. Soc.* **1968**, A, 3047.
(15) Deeming, A. J., Shaw, B. L., *J. Chem. Soc.* **1969**, A, 597.
(16) Eaton, D. R., Suart, S. R., *J. Am. Chem. Soc.* **1968**, 90, 4170.
(17) Emsley, J. W., Feeny, J., Sutcliffe, L. H., "High Resolution Nuclear Magnetic Resonance Spectroscopy," Vol. 2, p. 989, Pergamon, Oxford, 1966.
(18) Fackler, J. P., Jr., Fetchin, J. A., Seidel, W. C., *J. Am. Chem. Soc.* **1969**, 91, 1217.
(19) Fackler, J. P., Jr., Seidel, W. C., Fetchin, J. A., *J. Am. Chem. Soc.* **1968**, 90, 2707.
(20) Glass, G. E., Schwabacher, W. B., Tobias, R. S., *Inorg. Chem.* **1968**, 7, 2471.
(21) Glass, G. E., Tobias, R. S., *J. Am. Chem. Soc.* **1967**, 89, 6371.
(22) Griffith, J. S., Orgel, L. E., *Trans. Faraday Soc.* **1957**, 53, 601.
(23) Grim, S. O., Ference, R. A., *Inorg. Nucl. Chem. Letters* **1966**, 2, 205.
(24) Grim, S. O., Keiter, R. L., McFarlane, W., *Inorg. Chem.* **1967**, 6, 1133.
(25) Heaton, B. T., Pidcock, A., *J. Organometal. Chem.* **1968**, 14, 235.
(26) Holloway, C. E., Hulley, G., Johnson, B. F. G., Lewis, J., *J. Chem. Soc.* **1969**, A, 53.
(27) Jenkins, J. M., Shaw, B. L., *J. Chem. Soc.* **1966**, A, 770.
(28) *Ibid.*, **1966**, 1407.
(29) Jenkins, J. M., Shaw, B. L., *Proc. Chem. Soc.* **1963**, 279.
(30) Jensen, K. A., *Z. Anorg. Allgem. Chem.* **1936**, 229, 252.
(31) Kaplan, P. D., Schmidt, P., Orchin, M., *J. Am. Chem. Soc.* **1967**, 89, 4537.
(32) Kettle, S. F. A., *J. Chem. Soc.* **1965**, 6664.
(33) Kite, K., Smith, J. A. S., Wilkins, E. J., *J. Chem. Soc.* **1966**, A, 1744.
(34) Lydon, J. E., Truter, M. R., *J. Chem. Soc.* **1965**, 6899.
(35) McFarlane, W., *Chem. Commun.* **1967**, 772.

(36) *Ibid.*, **1968**, 393.
(37) *Ibid.*, **1969**, 439.
(38) *Ibid.*, **1969**, 700.
(39) McFarlane, W., *J. Chem. Soc.* **1967**, A, 1922.
(40) McFarlane, W., *Quart. Rev.* **1969**, 23, 187.
(41) Morgan, G. L., Rennick, R. D., Soong, C. C., *Inorg. Chem.* **1966**, 5, 372.
(42) O'Brien, J. F., Glass, G. E., Reynolds, W. L., *Inorg. Chem.* **1968**, 7, 1664.
(43) Osborn, J. A., Jardine, F. H., Young, J. F., Wilkinson, G., *J. Chem. Soc.* **1966**, A, 1711.
(44) Parshall, G. W., *J. Am. Chem. Soc.* **1964**, 86, 5367.
(45) *Ibid.*, **1966**, 88, 704.
(46) Pidcock, A., Richards, R. E., Venanzi, L. M., *J. Chem. Soc.* **1966**, A, 1707.
(47) *Ibid.*, **1968**, 1970.
(48) Pidcock, A., Richards, R. E., Venanzi, L. M., *Proc. Chem. Soc.* **1962**, 184.
(49) Pople, J. A., Santry, D. P., *Mol. Phys.* **1964**, 8, 1.
(50) Ramsey, N. F., *Phys. Rev.* **1950**, 78, 699.
(51) Smith, B. B., Sawyer, D. T., *Inorg. Chem.* **1969**, 8, 379.
(52) *Ibid.*, **1968**, 7, 922.
(53) *Ibid.*, **1968**, 7, 1526.
(54) Smith, J. A. S., *J. Chem. Soc.* **1962**, 4736.
(55) Thomas, S., Reynolds, W. L., *Inorg. Chem.* **1969**, 8, 1531.
(56) Tobias, R. S., *Inorg. Chem.* **1970**, 9, 1296.
(57) Varian Associates, "NMR Table," 5th ed., 1965.
(58) Walton, R. A., *J. Chem. Soc.* **1967**, A, 1852.
(59) Yagupsky, G., Wilkinson, G., *J. Chem. Soc.* **1969**, A, 725.

RECEIVED December 16, 1969.

9

Crystal Structures of Complexes of the Platinum Group Metals

E. L. AMMA

University of South Carolina, Columbia, S. C. 29208

> *The crystal and molecular structure of selected groups of Pt metal complexes are reviewed and discussed. These structures are divided into three groups: molecular adducts, "square planar" Pt(II) complexes, and structures involving delocalized π systems. The molecular adducts of N_2, O_2, CO, and NO^+ are classified (in general) into two types, either trigonal bipyramids or distorted tetragonal pyramids. The geometry and bond lengths of the Pt(II) complexes are discussed in terms of limitations imposed on discussions of bond lengths. The effect of π electrons on complex stereochemistry is discussed in the remaining section.*

This summary of the x-ray structure determinations of Pt group metals is not to be construed as comprehensive. Rather, it should be viewed as recent structure determinations of Pt group metals of interest to the author; any omissions should be considered as limitations in the scope of interest of the author and not in the importance of the research omitted.

The crystal structures of the Pt group metals to be examined fall readily into three categories: (1) New complexes where the stereochemistry about the metal atom is of interest, e.g., molecular adducts. These results are intrinsically of interest and importance to the nature of the chemical bonding in these molecules or ions. The inclusion of the structures of this class of compounds is particularly appropriate since their chemistry and kinetic properties are discussed elsewhere in this symposium. (2) The "accurate" determination of bond lengths in order to investigate π bonding or the nature of metal–ligand bonds in general. (3) Pt metal structures containing, or potentially containing, delocalized π systems. A recent review containing a number of Pt structures is: Churchill, M. R., "Structural Studies on Transition Metal Complexes Containing σ-Bonded Carbon Atoms" *in* "Perspectives in Structural Chemistry," J. D. Dunitz and J. A. Ibers, Eds., Wiley, New York, 1970.

It seems appropriate at this time to put forth some cautionary statements for the general reader concerning these heavy metal atom structures and their quoted accuracy in bond lengths. In general, crystallographers are well aware of the shortcomings of their data. The quoted errors in bond lengths are the results of mathematical treatment of the data assuming no uncorrected systematic errors. All crystallographic data, including modern counter data collected by automated diffractometers, contain uncorrected systematic errors even after corrections for absorption, anomalous dispersion, etc., have been made. Some of the inherently remaining errors are instrumental, involving the monochromaticity of the radiation, stability and homogeneity of the source, as well as the stability of the detector and associated electronic systems. Mechanical misalignments are not to be overlooked. Crystals themselves may undergo radiation damage during the data collection period, and even more mundane occurrences such as crystal movement can create serious difficulties for the unwary. The theoretical models used for the calculation of the x-ray scattering by electrons in molecules have their shortcomings as well. Thermal motions of atoms from x-ray intensity data of crystals are not well understood. (Thermal ellipsoids of atoms are now common in many x-ray structure publications; they should be interpreted and, for that matter, accepted with caution.) We suggest that the practicing chemist multiply the quoted error by two unless some qualifying statements have been made, and when making comparisons between bond lengths in different structures, not consider differences significant unless they are at least twice this inflated error. In cases where disorder or unusually large thermal motion is reported, one is well advised to be even more conservative. Prior to approximately five to six years ago, almost all data were collected by photographic techniques, and photographic data have even more innate systematic errors. Hence, any detailed discussion about small differences in bond lengths from data before and photographic data after that date should be made with caution. Nevertheless, there is no other method of structure determination with a comparable magnitude of reliability for the determination of structures of moderate to complicated molecules.

Molecular Adducts

O_2 **Adducts.** Ibers continues his excellent structural investigations of O_2 adducts which started with $[(C_6H_5)_3P]_2IrCOZ$ in which $Z = Cl$ (*37*) and I (*44*). The structure of the Br isomer has also been recently solved (*71*). These studies by Ibers *et al.* have been extended to include

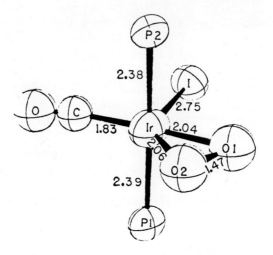

Inorganic Chemistry

Figure 1. The local environment about the Ir atom in $[(C_6H_5)_3P]_2$ Ir COCl · O_2

Bond length errors: Ir–P ± 0.008A, Ir–I ± 0.005A, Ir–O ± 0.02A, O–O ± 0.026A. Angle errors are 0.7° or less. (Ref. 44)

the oxygen adducts of $[(C_6H_5)_2P(CH_2)]_2Ir^+(PF_6^-)$ (45, 46), $[(C_6H_5)_2P(CH_2)]_2Rh^+(PF_6^-)$ (46). Other workers have solved the structure of the O_2 adduct of $[(C_6H_5)_2(C_2H_5)P]_2IrCOCl$ (77). Considering (for pictorial purposes) O_2 as a unidentate ligand, the structures of the first three and the very last complex can be described as trigonal bipyramidal with trans apical phosphines while CO and Z accompany the O_2 molecule in the trigonal plane (Figure 1). The geometry of the chelated diphos,- [bis(1,2-diphenylphosphino)ethane] Rh and Ir complexes may be described in an analogous manner.

Of considerable chemical and potentially of biological importance is the change in ease with which the O_2 molecule can be added to the parent complex as a function of ligands and central metal ion. The changes in bond lengths (O–O and M–O) have been correlated with the reversibility of oxygen uptake (46). The more electronegative the ligand P < I < Br < Cl, the less is the back donation of electrons into the O_2 ligand from the metal, and the O–O distance is more like free O_2. Likewise, the poorer the energy match between metal and oxygen orbitals, the more like free O_2 is the coordinated oxygen—*i.e.*, in $[(C_6H_5)_2P(CH_2)]_2RhO_2^+$, the O–O bond length is shorter than in $[(C_6H_5)_2P(CH_2)]_2IrO_2^+$. This explanation is in agreement with the facts, but the nature of the metal–oxygen bond is probably more complex and more sensitive to subtle changes at the metal than this simple picture indicates. For example, there is the variation in Ir–P distance of 0.17A in the $[(C_6H_5)_2$-$P(CH_2)]_2IrO_2^+$ (45, 46) complex, but no significant difference in Rh–P bond lengths in the analogous Rh adduct. Furthermore, the O–O distance of 1.63(2)A in this Ir–O_2 adduct is longer than the O_2 distance of 1.48(2)A in H_2O_2 (76). However, this distance may be understood if one considers mixing of excited states of O_2 rather than the ionic formulation O_2^{2-}. In addition, the O–O distance is surprisingly sensitive to changes in the phosphine going from 1.30(3)A in $[(C_6H_5)_3P]_2IrCOCl \cdot O_2$ (37) to 1.54(4)A in $[(C_6H_5)_2(C_2H_5)P]_2IrCOCl \cdot O_2$ (77). In relation to these structures, the structure of $[(C_6H_5)_3P]_2PtO_2$ (34) is interesting. In this case the metal, two phosphorus, and two oxygen atoms lie in the same plane accompanied by Pt–O distances of 2.01(3)A and an O–O distance of 1.45(4)A. This Pt–O distance is within the range of Ir–O distances found by Ibers, but the O–O distance is essentially that expected for O_2^{2-}. Since this complex does not in any sense of the term reversibly add O_2 and the geometry is "square planar," it should be described as Pt^{2+} with O_2^{2-}. Support for this formulation is found in the structure of tris(triphenylphosphine)carbonyl platinum which may be described as tetrahedral (1)—i.e., Pt(O).

N_2 **Adducts.** The crystal structures of three nitrogen adducts have been solved and reported at this time. Ibers and coworkers have determined the structures of $CoH(N_2)[P(C_6H_5)_3]_3$ (16) and $\{Ru(N_3)(N_2)$-$[NH_2(CH_2)_2NH_2]_2\}^+PF_6^-$ (15). The environment about the cobalt atom is trigonal bipyramidal with the three phosphorus atoms in the trigonal plane, whereas the N_2 molecule and the hydride are in the apical positions. In contrast, the Ru is six-coordinate. The M–N–N angle in both cases is 180°, and the N_2 molecule is bound end-on to the metal, as is the CO moiety in carbonyl complexes. The Co–N bond length of 1.80(1)A av. also denotes some metal–nitrogen multiple bonding. The structure of $Ru(NH_3)_5N_2Cl_2$ was solved by Nyburg and Bottomly (13), and they also report a six-coordinate Ru with a linear Ru–N–N arrangement. However, this structure is complicated by disorder.

SO_2 **and** CS_2 **Adducts.** Ibers *et al.* have published the structures of the SO_2 adducts of $[(C_6H_5)_3P]_2M$–COCl in which M = Ir (36) or Rh (53). The stereochemistry of the metal atoms is that of a tetragonal pyramid with CO, Cl, and trans P atoms in the basal plane and S of the SO_2 group at the apex. The M–S vector in these compounds makes an angle of $\sim$ 30° with the normal to the SO_2 plane. Note the difference between this geometry and that of the analogous O_2 adducts. Vogt, Katz, and Wiberly (80) reported the crystal structure of $[RuCl(NH_3)_4SO_2]Cl$ wherein the Ru is "octahedrally" coordinated with Cl trans to S. The S of the SO_2 group is tilted similarly to that found in the above-mentioned Ir and Rh complexes.

The crystal structure of the carbon disulfide adduct of $[(C_6H_5)_3P]_3Pt$ has been solved (*42*). In this case, the Pt is bound to a carbon and one sulfur of the CS_2 group. The Pt atom and its four directly-bound neighbors are coplanar, but the unbound sulfur is tipped out of this plane. Furthermore, the CS_2 molecule is bent with a S–C–S angle of 136(4)°. It is an interesting question in this case as to the oxidation state of the metal relative to the geometry of the complex.

Tetracyanoethylene Adducts (TCNE). The structure of the tetracyanoethylene adduct (*45*) of $[(C_6H_5)_3P]_2IrCOBr$ has been determined, and in contrast to the O_2 adduct, although the stereochemistry about Ir is also trigonal bipyramidal with the C=C in the equatorial plane, the phosphines are cis and in the equatorial plane. The C=C distance in the adduct is 0.18A longer than in free tetracyanoethylene and the TCNE entity is nonplanar. These facts can be readily interpreted as the Ir donating charge into the π^* orbital of TCNE. The structure of $Ir(C_6N_4H)$-$CO(TCNE)[P(C_6H_5)_3]_2$ (*60*) is very similar to the above except that the axial Ir–Br interaction is replaced by an Ir–N sigma bond from a modified TCNE ligand. In contrast, the structure of the $Pt[P(C_6H_5)_3]_2$-TCNE (*12*) complex yields an approximately planar PtP_2C_2 unit and again, as in the CS_2 adduct of Pt(O), an unusual stereochemistry.

Nitrosyl Adducts. Three structures can be related to the $IrCOCl[P-(C_6H_5)_3]_2 \cdot SO_2$ structure directly—*i.e.*, they have a tetragonal pyramid structure with a bent axial M–N–O grouping. These are: $\{IrI(CO)(NO)-[P(C_6H_5)_3]_2\}BF_4 \cdot C_6H_6$ (*31*), $\{IrCl(CO)(NO)[P(C_6H_5)_3]_2\}BF_4$ (*30*, *32*), and $RuCl(NO)_2[P(C_6H_5)_3]_2PF_6$ (*59*). In the latter structure, the only significant difference is that one of the NO groups is in the basal plane with a linear Ru–N–O grouping similar to the CO of the SO_2 complex. A linear M–N–O arrangement was also found in "octahedral" $OsCl_2(HgCl)NO[P(C_6H_5)_2]_2$ (*7*). On the other hand, the structure of $Ir(NO)_2[P(C_6H_5)_3]_2ClO_4$ (*52*) consists of a distorted tetrahedron of the nearest neighbor atoms about Ir with almost linear Ir–N–O groupings.

CO Adducts. This is somewhat unusual terminology, but we restrict this brief discussion to molecules related to the above. The structures of $IrCl(CO)_2[P(C_6H_5)_3]_2 \cdot C_6H_6$ (*58*) and $Os(CO)_3[P(C_6H_5)_3]_2$ (*73*) have been determined, and they are trigonal bipyramidal, as is the analogous O_2 analog of the former of the two mentioned herein except that CO is bound end-on in both cases.

Stereochemistry and Bond Lengths in Pt(II) Complexes

Basolo and Pearson (*5b*) as part of a discussion of the kinetic trans effect (for definition see Ref. *5a*) established a qualitative trans influence series by noting the Pt–X bond length as a function of the trans partner L in L–Pt–X. The resulting series is quite similar to the kinetic trans effect series. However, it is now generally accepted that the trans effect is an outdated concept when viewed in terms of the details of the electronic structure of the complex and activated intermediates (35, 85).

The data on bond lengths used in the series by Basolo and Pearson are interesting examples of older data that have been used to draw conclusions. (This is not to find fault with Basolo and Pearson; that's all they had!) A number of these structures have been redetermined by modern methods and the earlier work was found to be unreliable. Furthermore, the use of the term trans-influence is still to be found in the literature as a bond length feature.

The most striking example of the difference between earlier work (11, 47, 48) and the more recent structure (9, 33) refinement is that of Zeise's salt (KPtCl$_3$ · C$_2$H$_4$ · H$_2$O) (Figure 2). The older results showed a Pt–Cl bond length of 2.42A for Cl trans to C$_2$H$_4$ and a 2.32A Pt–Cl bond length for Cl trans to Cl. As can be seen in Figure 2, there are no significant differences between the three Pt–Cl distances. A "normal" Pt–Cl single bond length may be taken as 2.30A (41, 49, 57). This particular compound was used for some time as the example for π-bonding and trans influence! The Pt–Cl bond length trans to ethylene in dipentene Pt(II) chloride (4) is 2.33(1)A, but is rarely referenced. The term trans influ-

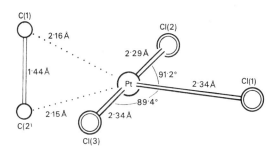

Acta Crystallographica

Figure 2. The local environment about the Pt atom in Zeise's salt [KPtCl$_3$C$_2$H$_4$ · H$_2$O]

Bond length errors: Pt–Cl(1) ± 0.004A, Pt–Cl(2), Cl(3) ± 0.020A, Pt–C ± 0.020A. Angle errors are 0.6° or less. (Ref. 9)

A more recent refinement (private communication from P. G. Owston) makes the distances as follows:
Pt–Cl(1) 2.327(5) A
Pt–Cl(2) 2.314(7) A
Pt–Cl(3) 2.296(7) A

ence relative to bond lengths continues in use but one must bear in mind that for small changes in bond lengths a multitude of factors can be responsible, such as ionic forces if the complex is ionic, hydrogen bonding, and molecular packing, to name a few. Intermolecular forces and their effect on bond lengths and angles are minimized for uncharged complexes, and the most reliable distances for comparison purposes can be obtained from these systems. However, even in these cases, caution must be exercised; for example, the structure of trans-dichlorobistripropylphosphine-u,u'-dichloroplatinum (10) in which the center of the chlorine-bridged dimer lies on a center of symmetry. The Pt–Cl distance of the bridging chlorine trans to the phosphine group is long at 2.425(8)A and the Pt–P distance is "short" at 2.230(9)A. A Pt–P distance of 2.247(7)A is observed in cis-Pt[P(CH$_3$)$_3$]$_2$Cl$_2$ (50). The elongation of the Pt–Cl distance from the normal 2.30A is attributable to two factors: (a) Inevitably, the Pt–Cl distance with the bridging chlorine will be long because of the halogen forming two bonds and consequently, bearing an effective formal positive charge and (b) the fact that it is trans to the phosphine (Pt–Cl in cis-Pt[P(CH$_3$)$_2$]$_2$Cl$_2$ is 2.38A). The terminal Pt–Cl distance is normal at 2.279(9)A within two standard deviations. Therefore, care must be exercised in terms of how much bond lengthening of the Pt–Cl bond is attributable to the phosphine alone. The structures of cis- and trans-dichlorodiamminoplatinum(II) determined by Milburn and Truter (51) are the most reliable structure determinations available of simple Pt(II) chloroammines. The Pt–N distances are 2.03(4)Av., whereas the Pt–Cl distances are 2.32(1)Av. and are not significantly different in the two isomers. Essentially the same results for Pt–N distances were found in trans-bis(dimethylamine)diamine platinum(II) chloride (3). Recent structures containing Pt–N distances are bis(2-amino-2-methyl-3-butanone-oximato)platinum(II) chloride monohydrate (64) and bis(glyoximato)platinum(II) (23). There is little doubt about the bond lengthening that occurs for Pt–X trans to hydride, as has been shown in Pt[P(C$_6$H$_5$)$_3$]$_2$HBr (56) and Pt[P(C$_6$H$_5$)$_2$C$_2$H$_5$]$_2$(H)Cl (20), where the lengthening is approximately 0.1A in both cases.

Turning to other ligands that have strong trans influence, the structure of dichlorodi(o-phenylenebisdimethylarsine)platinum(II) (74) which is "square planar" with Pt–As bonds of 2.37(1)A is considerably shorter than the 2.49A (57) expected from the sum of single-bond radii. Unfortunately, there are no structure data on cis-PtCl$_2$(AsR$_3$)$_2$ for comparison purposes, but the bridging chlorine trans to As(CH$_3$)$_3$ has a Pt–Cl distance in di-u-chloro-[dichloro-bis(trimethylarsine)]platinum(II) longer at 2.39A than the bridging chlorine Pt–Cl distance trans to Cl at 2.31A. The Pt–As distance is concomitantly 2.31A (somewhat shorter than for the As–Pt–As interaction).

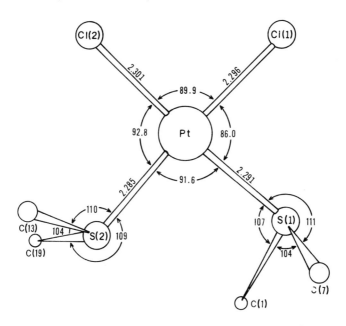

Figure 3. The local environment about the Pt atom in cis-$[(p\text{-}Cl\text{-}C_6H_4)_2 \cdot S]_2 PtCl_2$

Bond length errors are: Pt–S = Pt–Cl ± 0.007A, S–C = C–C = ±0.02A. Angle errors are ±0.6° or less. S–C distances are not significantly different from 1.80A.

Sulfur is generally considered as having weaker trans influence than phosphines; however, thiourea has a greater trans influence than thioether. Very few structures of Pt or Pd thioether complexes have been solved. Woodward et al. (28, 62) have shown that in $(R_2S)_2M_2Br_4$ when M is Pt, the complex is a sulfur-bridged dimer with relatively short Pt–S bridging distances at 2.209(7) and 2.242(14)A, and Pt–Br terminal distances of 2.384(4)A and 2.400(7)A. When M is Pd, X = Br, the complex is a trans Br-bridged dimer with Pd–Br bridging distances of 2.429(4)A and 2.447(11)A, Pd–Br terminal distances of 2.404(4)A, and Pd–S distances of 2.30(2)A. Interesting as the contrast in these two structures may be, they say little about the trans influence of the Pt–S bond on the Pt–X bond length. In part to examine this question, we have solved the structure of dichlorobis(4,4'-dichlorodiphenylsulfide)platinum(II) (70) (Figure 3). The 2S, 2Cl, and Pt atoms are essentially in the same plane, and the PtS_2Cl_2 unit may be described as "square planar." The Pt–Cl distances are normal single bonds, as are the Pt–S distances. There is no elongation of the Pt–Cl distances owing to a trans thioether. The Pt–S length is only slightly shorter than the 2.35A expected from the sum of the covalent radii. The C–S–Pt angles are such that it is clear that S sp^3

orbitals are used to form the Pt–S bond. Further, the orientation of the p-chlorophenyl rings is such that no significant π interaction exists between the rings and the sulfur or Pt atoms. Hence, the trans influence of a thioether is not reflected in any Pt–Cl bond elongation. A means of testing for the influence of π-type interactions in Pt–S bonds would be to use a ligand with a geometry that can be precisely located relative to the metal, and the location of the π orbitals in this ligand is unambiguous. A ligand that fulfills these criteria is thiourea (tu); the molecule is planar in all known complexes with metals, and the π orbitals are normal to the molecular plane. Further, the relative energies of the π levels in this molecule have been calculated. A simple MO calculation of the energy levels gives a strongly bonding $a_1(-2.23\beta)$, moderately bonding b_1 (-1.50β), weakly bonding $a_1'(-0.81\beta)$, and strongly antibonding a_1'' $(+1.03\beta)$ (2). With six electrons (one electron from S, C, and two from

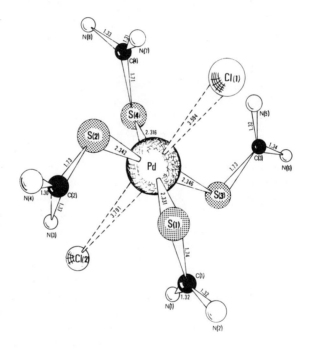

Figure 4. The molecular structure of $Pd[SC-(NH_2)_2]_4Cl_2$ showing the principal interatomic distances

Bond length errors are: Pd–S = Pd–Cl ± 0.003A, S–C ± 0.011A, C–N ± 0.014A. For clarity the angles are omitted. The angles are: S–Pd–S approximately 90° with errors of 0.1°, Pd–S–C ~ 110° with errors of 0.4°, S–C–N ~ 120° with errors of 1° or less. N–C–N ~ 120° with errors of 1° or less.

each N), these levels are filled through a_1'. Although this molecule can behave as a π donor (69, 72, 78, 79, 81) without sigma bonding, it has been shown by others (14, 25, 26, 38, 40, 61) as well as ourselves (8, 54, 55, 84) that, in general, it forms bonds with transition metals via the sp^2 orbitals of sulfur. Sterically, it is possible to construct a model such that all four thiourea groups are coplanar with the metal—i.e., with C_{4h} symmetry. Although we have detailed x-ray structure data on $Pttu_4Cl_2$, $Pdtu_4Cl_2$ (8), $Nitu_4Cl_2$ (38), $Cotu_4Cl_2$ (54, 55), and $Nitu_6Br_2$ (84), the general important features can be seen from $Pdtu_4Cl_2$ (Figure 4). The $Pdtu_4^{2+}$ unit is not planar, but rather the metal (neglecting the Cl atoms) defines an approximate molecular center of symmetry and the thiourea groups are tipped relative to the approximate PdS_4 plane by 43°–60° and twisted relative to the Pd–S–C lines by 17°–26°. The tilt is defined as the dihedral angle between planes—e.g., Pd S(1), S(3) and S(1), S(3), C(1). The twist is defined as the dihedral angle between planes—e.g., Pd S(1), C(1) and S(1), C(1), N(1), N(2). It is tempting to say that these orientations are governed by hydrogen bonding and packing considerations. However, approximately the same orientations are found for the thiourea groups independent of metal coordination number and anion. In lieu of a complete description of the electronic structure of the complex, a possible but not unique interpretation of these results is that it is energetically too expensive to remove electrons from the metal $d\pi$ orbitals to the tu π a_1'' MO. The system distorts (tilts and twists) in order to make use of empty nonbonding $3d$ orbitals on the sulfur atom.

A further test of this hypothesis can be made by changing the energy levels of the ligand by, e.g., going to thioacetamide (tac) with three π levels at energies (-1.96β), (-0.91β), and ($+0.86\beta$). In a similar manner, the π^* level of thiourea may be made more accessible by replacing hydrogen atoms by electron donor substituents, e.g., sym-dimethylthiourea (DMT). The crystal structures of $Nitac_4Br_2$ and $Ni(DMT)_4Br_2$ have indeed significantly different structural features from $Pd(tu)_4^{2+}$, and these differences can be related to the discussion above.

Structures of Delocalized π-Systems Involving Pt Metals

Much interest has been generated in various properties of metal derivatives of dithioketones, ethylene (1, 2) dithiolates, and related compounds in recent years (17, 18, 29, 43, 65, 66). There are many interesting facets of these fascinating compounds, and the previously mentioned references cover them in some detail. We limit ourselves here to the structures of the Pt group metals of these engrossing ligands. Unfortunately, some of the more interesting properties of the transition metals

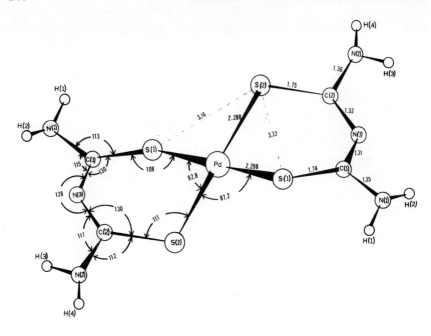

Figure 5. The molecular structure of bis(dithiobiureto)Pd(II), Pd(dtb)$_2$; the structural parameters and errors are listed in Table 1 with the analogous parameters for Ni(dtb)$_2$

Table I. Comparison of the Structural

Function	Pd(dtb)$_2$	Ni(dtb)$_2$
Ligand charge	−1	−1
Chair angle[b]	38°	11°
Maximum deviation from planarity[d]	0.51A	0.26A
Distances		
a. Interligand S–S	3.169(2)	2.895(6)
b. Intraligand S–S	3.319(2)	3.220(6)
c. Metal–S	2.301(1)	2.160(2)
	2.288(1)	2.171(2)
d. S–C	1.712(7)	1.720(8)
	1.743(7)	1.728(8)
e. C–N	1.32(1)	1.32(1)
	1.34(1)	1.34(1)
	1.35(1)	1.34(1)

[a] Tilt: Angle between planes defined by: M, S(1), S(2) and S(1), C(1), C(1) or M, (S(2), C(2).
[b] Chair Angle: Angle between planes defined by M, S(1), S(2) and S(1), S(2), N(3).

with these ligands are not manifested with the Pt metals. It has been shown that the structures:

$$\begin{bmatrix} R & S & S & R \\ \diagdown C \diagup & \diagdown & \diagup C \diagdown \\ | & M & | \\ \diagup C \diagdown & \diagup & \diagdown C \diagup \\ R & S & S & R \end{bmatrix}_n$$

are completely planar for R = C_6H_6, M = Ni, n = 0 (63); R = CN, M = Ni, n = −2 (19); R = CN, M = Ni, n = −1 (24). It is generally accepted that for M = Pt, Pd, they are isostructural with the Ni analog. Further, it has been shown that expansion of the chelate ring also yields planar structures, for example, bis(dithioacetylacetone)Ni(II) and Co(II) (6).

We decided to examine what happens structurally to the complex upon going to a six-membered ring and in addition, adding an excess of nonbonding π electrons from a planar ligand. Presumably, the expansion to a six-membered ring does not effect the planarity, but the addition of π electrons might. We chose the ligand dithiobiuret ($H_2NCSNHCSNH_2$). We found that this ligand reacts as a negative anion with loss of the proton from the central nitrogen, forming neutral complexes M(dtb)$_2$ with Pd, Pt, and Ni. This ligand has four more π electrons than does

Parameters of Ni(dtb)$_2$ and Pd(dtb)$_2$

Functions	Pd(dtb)$_2$	Ni(dtb)$_2$
Tilt[a]	41.1°	3.23°
	30.1°	18.6°
Twist[c]	32.2°	23.4°
	4.4°	12.4°
Angles		
a. M–S–C	108.8(2)	116.4(2)
	111.6(2)	115.3(2)
b. S–M–S	87.4(1)	83.9(1)
(Interligand)		
c. S–C–N	131.1(5)	130.8(7)
(Interior)	131.3(5)	130.2(7)
d. C–N–C	126.0(6)	125.0(8)

[c] Twist: Angle between planes defined by: M, S(1), C(1) and S(1), C(1) N(1), N(3); M, S(2), C(2) and S(2), C(2) N(2) N(3).
[d] The S(1), C(1), N(1), N(3) and S(2), C(2), N(2), N(3) units are rigorously planar.

dithioacetylacetone. We have determined the crystal structures of these three complexes (*27, 39*). The structure of the Pd(dtb)$_2$ is shown in Figure 5. Although the Ni, Pt, and Pd structures are similar, there are nevertheless some important differences; these are summarized in Table I. In each case, the metal and its four sulfur atoms are planar, but the entire molecule is distinctly not planar; in fact, it is in a chair form. In changing the metal from Pd to Ni, the molecule becomes more planar. If nonbonded repulsions between the sulfur atoms on opposite ligands were responsible for the nonplanarity, it would be expected that the Ni complex would be less planar because the shorter Ni–S distance pulls the nonbonded sulfur atoms closer together. This is clearly not the case. The Pd, Pt complexes could be more distorted from planarity because of steric strain introduced into the six-membered ring by the longer (Pd)Pt–S distances. Alternatively, the distortion could arise from the fact that the extra π electrons on the nitrogen would go into antibonding MO's if the entire molecule were planar and since, in general, there is more metal–ligand π interaction with Pd or Pt than Ni, it would be expected that these two complexes would be less planar. Another explanation that might be advanced for this change in geometry is that the action of pulling the sulfur atoms closer together brings about some bonded S–S interaction. However, this type of interaction has not been proven as a stabilization factor. This S–S interaction has been suggested (*67, 75*) as the reason for trigonal prismatic coordination in M(S$_2$C$_2$R$_2$)$_3$ (*21, 22, 68*). Even though there seems to be a S–S π–π interaction between the individual S$_2$C$_2$R$_2$ groups, this interaction alone would not stabilize trigonal prismatic coordination so long as the ligands are neutral or dianions.

Acknowledgment

The authors received financial support from National Institutes of Health, Grants GM-13985 and HE-12523.

Literature Cited

(1) Albano, V. G., Ricci, G. M. B., Bellon, P. L., *Inorg. Chem.* **1969**, 8, 2109.
(2) Amma, E. L., unpublished calculations.
(3) Anderson, J. S., Carmichael, J. W., Cordes, A. W., *Inorg. Chem.* **1970**, 9, 143.
(4) Baenziger, N. C., Medrud, R. C., Doyle, J. R., *Acta Cryst.* **1965**, 18, 237.
(5) Basolo, F., Pearson, R. G., "Mechanisms of Inorganic Reactions," 2nd ed., a) p. 355, b) p. 360, Wiley, New York, 1967.
(6) Beckett, R., Hoskins, B. F., *Chem. Commun.* **1967**, 909.
(7) Bentley, G. A., Laing, K. R., Roper, W. R., Waters, J. M., *Chem. Commun.* **1970**, 998.

(8) Berta, D. A., Spofford, W. A., III, Boldrini, P., Amma, E. L., *Inorg. Chem.* **1970**, 9, 136.
(9) Black, M., Mais, R. H. B., Owston, P. G., *Acta Cryst.* **1969**, B25, 1753.
(10) *Ibid.*, **1969**, B25, 1760.
(11) Bokii, G. B., Kukina, G. A., *Zh. Strukt. Khim.* **1965**, 5, 706.
(12) Bombieri, G., Forsellini, E., Panattoni, G., Graziani, R., Bandoli, G., *J. Chem. Soc.* **1970**, A 1313.
(13) Bottomley, F., Nyburg, S. C., *Chem. Commun.* **1966**, 897.
(14) Capacchi, L., Gasparri, G. F., Nardelli, M., Pelizzi, G., *Acta Cryst.* **1968**, B24, 1199.
(15) Davis, B. R., Ibers, J. A., *Am. Cryst. Assoc. Meeting, New Orleans, La., March 1970*, Abstr. No. C-9; *Inorg. Chem.* **1970**, 9, 2768.
(16) Davis, B. R., Payne, N. C., Ibers, J. A., *Inorg. Chem.* **1969**, 8, 2719.
(17) Davison, A., Edelstein, N., Holm, R. H., Maki, A. H., *J. Am. Chem. Soc.* **1963**, 85, 2029.
(18) *Ibid.*, **1964**, 86, 2799.
(19) Eisenberg, R., Ibers, J. A., *Inorg. Chem.* **1965**, 4, 605.
(20) *Ibid.*, **1965**, 4, 773.
(21) Eisenberg, R., Ibers, J. A., *J. Am. Chem. Soc.* **1965**, 87, 3776.
(22) Eisenberg, R., Stiefel, E. I., Rosenberg, R. C., Gray, H. B., *J. Am. Chem. Soc.* **1966**, 88, 2874.
(23) Ferraris, G., Viterbo, D., *Acta Cryst.* **1969**, B25, 2066.
(24) Fritchie, C. J., Jr., *Acta Cryst.* **1966**, 20, 107.
(25) Gasparri, G. F., Mangia, A., Musatti, A., Nardelli, M., *Acta Cryst.* **1969**, B25, 203.
(26) Gasparri, G. F., Musatti, A., Nardelli, M., *Chem. Commun.* **1966**, 602.
(27) Girling, R. L., Amma, E. L., *Chem. Commun.* **1968**, 1487.
(28) Goggin, P. L., Goodfellow, R. J., Sales, D. L., Stokes, J., Woodward, P., *Chem. Commun.* **1968**, 31.
(29) Gray, H. B., *Trans. Metal Chem.* **1965**, 1, 239.
(30) Hodgson, D. J., Ibers, J. A., *Inorg. Chem.* **1968**, 7, 2345.
(31) *Ibid.*, **1969**, 8, 1282.
(32) Hodgson, D. J., Payne, N. C., McGinnety, J. A., Pearson, R. G., Ibers, J. A., *J. Am. Chem. Soc.* **1968**, 90, 4486.
(33) Jarvis, J. A. J., Kilbourn, B. T., Owston, P. G., *Acta Cryst.* **1970**, B26, 876.
(34) Kashiwagi, T., Yasuoka, N., Kasai, N., Kakudo, M., Takahashi, S., Hagihara, N., *Chem. Commun.* **1969**, 743.
(35) Langford, C. H., Gray, H. B., "Ligand Substitution Processes," Benjamin, New York, 1966.
(36) La Placa, S. J., Ibers, J. A., *Inorg. Chem.* **1966**, 5, 405.
(37) La Placa, S. J., Ibers, J. A., *J. Am. Chem. Soc.* **1965**, 87, 2581.
(38) Lopez-Castro, A., Truter, M. R., *J. Chem. Soc.* **1963**, 1309.
(39) Luth, H., Hall, E. A., Spofford, W. A., III, Amma, E. L., *Chem. Commun.* **1969**, 520.
(40) Luth, H., Truter, M. R., *J. Chem. Soc.* **1968**, A, 1879.
(41) Mais, R. H. B., Owston, P. G., Wood, A. M., unpublished results, 1968.
(42) Mason, R., Rae, A. I. M., *J. Chem. Soc.* **1970**, 1767.
(43) McCleverty, J. A., *Progr. Inorg. Chem.* **1968**, 10, 49.
(44) McGinnety, J. A., Doedens, R. J., Ibers, J. A., *Inorg. Chem.* **1967**, 6, 2243.
(45) McGinnety, J. A., Ibers, J. A., *Chem. Commun.* **1968**, 235.
(46) McGinnety, J. A., Payne, N. C., Ibers, J. A., *J. Am. Chem. Soc.* **1969**, 91, 6301.
(47) Mellor, D. P., Wunderlich, J. A., *Acta Cryst.* **1954**, 7, 130.
(48) *Ibid.*, **1955**, 8, 57.
(49) Messmer, G. G., Amma, E. L., *Inorg. Chem.* **1966**, 5, 1775.
(50) Messmer, G. G., Amma, E. L., Ibers, J. A., *Inorg. Chem.* **1967**, 6, 725.
(51) Milburn, G. H. W., Truter, M. R., *J. Chem. Soc.* **1966**, A, 1609.

(52) Mingos, D. M. P., Ibers, J. A., *Inorg. Chem.* **1970**, 9, 1105.
(53) Muir, K. W., Ibers, J. A., *Inorg. Chem.* **1969**, 8, 1921.
(54) O'Connor, J. E., Amma, E. L., *Chem. Commun.* **1968**, 892.
(55) O'Connor, J. E., Amma, E. L., *Inorg. Chem.* **1969**, 8, 2367.
(56) Owston, P. G., Partridge, J. M., Rowe, J. M., *Acta Cryst.* **1960**, 13, 246.
(57) Pauling, L., "Nature of the Chemical Bond," 3rd ed., p. 249, Cornell University Press, 1960.
(58) Payne, N. C., Ibers, J. A., *Inorg. Chem.* **1969**, 8, 2714.
(59) Pierpont, C. G., Van Derveer, D. G., Durlard, W., Eisenberg, R., *J. Am. Chem. Soc.* **1970**, 92, 4760.
(60) Ricci, J. S., Ibers, J. A., Fraser, M. S., Baddley, W. H., *J. Am. Chem. Soc.* **1970**, 92, 3489.
(61) Robinson, W. T., Holt, S. L., Jr., Carpenter, G. B., *Inorg. Chem.* **1967**, 6, 605.
(62) Sales, D. L., Stokes, J., Woodward, P., *J. Chem. Soc.* **1968**, A, 1852.
(63) Sartain, D., Truter, M. R., *Chem. Commun.* **1966**, 382; *J. Chem. Soc. A* **1967**, 1264.
(64) Schlemper, E. O., *Inorg. Chem.* **1969**, 8, 2740.
(65) Schrauzer, G. N., *Acct. Chem. Res.* **1969**, 2, 72.
(66) Schrauzer, G. N., *Trans. Metal Chem.* **1968**, 4, 299.
(67) Schrauzer, G. N., Mayweg, V. P., *J. Am. Chem. Soc.* **1966**, 88, 3235.
(68) Smith, A. E., Schrauzer, G. N., Mayweg, B. P., Heinrich, W., *J. Am. Chem. Soc.* **1965**, 87, 5798.
(69) Spofford, W. A., III, Amma, E. L., *Chem. Commun.* **1968**, 405.
(70) Spofford, W. A., III, Amma, E. L., *Natl. Mtg. ACS, 157th, Minneapolis, Minnesota, April 1969*, Abstr. INOR 133; *Inorg. Chem.*, in press.
(71) Spofford, W. A., III, Amma, E. L., to be published.
(72) Spofford, W. A., III, Boldrini, P., Amma, E. L., Carfagno, P., Gentile, P. S., *Chem. Commun.* **1970**, 40.
(73) Stalick, J. K., Ibers, J. A., *Inorg. Chem.* **1969**, 8, 419.
(74) Stephenson, N. C., *Acta Cryst.* **1964**, 17, 1517.
(75) Stiefel, E. I., Eisenberg, R., Rosenberg, R. C., Gray, H. B., *J. Am. Chem. Soc.* **1966**, 88, 2956.
(76) Sutton, L. E., Ed., "Tables of Inter-Atomic Distances and Configuration in Molecules and Ions," Supplement, 1956–59, Special Publication No. 18, p. S-9-s, The Chemical Society, London, 1965.
(77) Taylor, I. F., Jr., Weininger, M. S., Amma, E. L., to be published.
(78) Vizzini, E. A., Amma, E. L., *J. Am. Chem. Soc.* **1966**, 88, 2872.
(79) Vizzini, E. A., Taylor, I. F., Jr., Amma, E. L., *Inorg. Chem.* **1968**, 7, 1351.
(80) Vogt, L. H., Jr., Katz, J. L., Wiberly, S. E., *Inorg. Chem.* **1965**, 4, 1157.
(81) Vranka, R. G., Amma, E. L., *J. Am. Chem. Soc.* **1966**, 88, 4270.
(82) Watkins, S. F., *Chem. Commun.* **1968**, 504.
(83) Watkins, S. F., *J. Chem. Soc.* **1970**, A, 168.
(84) Weininger, M. S., O'Connor, J. E., Amma, E. L., *Inorg. Chem.* **1969**, 8, 424.
(85) Zumdahl, S. S., Drago, R. S., *J. Am. Chem. Soc.* **1968**, 90, 6669.

RECEIVED February 18, 1970.

10

^{195}Pt—A Survey of Mossbauer Spectroscopy

N. BENCZER-KOLLER

Department of Physics, Rutgers University, New Brunswick, N. J. 08903

> *The recoilless resonance absorption and scattering of the 99 keV transition in ^{195}Pt has been studied to determine the lifetime [$\tau = (1.7 \pm 0.2) \times 10^{-10}$ sec], conversion coefficient ($\alpha = 7.0 \pm 0.3$), magnetic moment ($\mu = -0.615 \pm 0.041$ nm), and radius relative to the ground state radius ($\delta R/R < 0$) of the first excited state of ^{195}Pt. The magnetic hyperfine structure of iron, cobalt, and nickel alloys was investigated. The internal field at the platinum nucleus in all these alloys was negative, and its magnitude suggests that its main contribution comes from polarization of conduction electrons. The temperature dependence of the Debye-Waller factor is in excellent agreement with predictions based on the quasiharmonic model.*

An interesting candidate for Mossbauer effect studies from the point of view of both nuclear structure and solid state investigations is ^{195}Pt. ^{195}Au decays to ^{195}Pt (*10*) (Figure 1) with a half-life of 192 days, and feeds two states at 99 and 130 keV, respectively, both of which in principle may be used for recoilless resonance experiments. However, both states have short mean lives and therefore relatively broad resonance lines. In addition, because of the high energy of the transitions, the recoilless fraction is small. Most of the results described below have been obtained from the study of the Mossbauer effect of the 99 keV transition alone. The mean life of the 99 keV first excited state, its magnetic moment, and the internal conversion coefficient α of the radiative transition are the nuclear properties of the state that were determined by Mossbauer spectroscopy. The interest in the solid state properties of platinum metal lies in the fact that magnetic properties of platinum alloys are nearly unknown compared with those of the iron group metals and alloys. Furthermore, the study of platinum metal lattice dynamics is amenable to straightforward theoretical analysis because of the simple crystal struc-

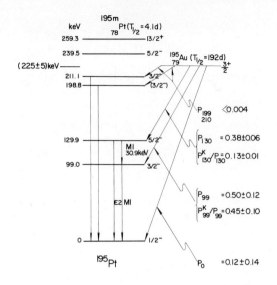

Figure 1. Decay scheme of ^{195}Au and energy level diagram of ^{195}Pt (10). P_i represents the relative transition probability for electron capture decay to the state i in ^{195}Pt and P_i^K/P_i is the ratio of K to total capture for decay to the state i.

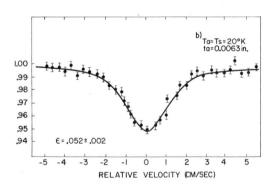

Figure 2. Typical Mossbauer absorption spectrum obtained with a ^{195}Au in a Pt matrix source and a 0.0063-inch thick Pt absorber; both source and absorber were at 20°K (9)

ture and vast available data on heat capacity, lattice constants, and elastic constants.

In a basic Mossbauer experiment, the reduction in transmission (9) (Figure 2) or the increase in scattered intensity of radiation (2) (Figure 3) is observed as a function of the relative velocity between a source and an absorber. The full width at half maximum of the resonance curve Γ is related to the mean life of the radiating state by the uncertainty relation $\Gamma \cong 2\hbar/\tau$. The depth of the curve, ϵ, is related to f, the magnitude of the recoilless fraction of gamma rays emitted, and hence to the crystalline properties of the solid. Finally, the displacement of the curve from zero relative velocity indicates the energy difference between emitted and absorbed radiation and is proportional to the s-electron

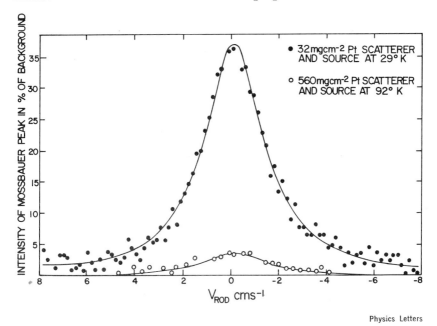

Figure 3. Typical Mossbauer scattering spectrum obtained with 32 mg/cm² Pt scatterer at 29°K and 560 mg/cm² Pt scatterer at 92°K (2)

density difference at the nucleus in the source and absorber, and to the difference in the mean square radius of the nucleus in its excited state and its ground state.

Resonance Effect and Line Width of the 99 and 130 keV Transitions

The 99 keV State. Typically, a source of 1–3 mCi of ^{195}Au electroplated on a platinum foil may be used. The resonance effect for the

99 keV transition for such a source and a 0.006-inch natural platinum foil absorber, both at 20°K, is $\epsilon \cong 5\%$ and the line width $\Gamma = 2.1$ cm/sec. The width of the resonant line depends on the absorber thickness, t_a. The extrapolated line width at $t_a = 0$, $\Gamma = (1.75 \pm 0.18)$ cm/sec, yields the half life of the state, $\tau_{1/2} = (1.7 \pm 0.2) \times 10^{-10}$ sec, if one assumes no broadening of the line owing to hyperfine interactions. This value agrees with the electronically measured value $\tau_{1/2} \approx 1.4 \times 10^{-10}$ sec (4). As platinum metal is a cubic crystal, no line broadening interactions are expected (9).

The 130 keV State. The decay of the 130 keV state has been studied extensively, and several inconsistencies are being resolved. The results of different measurements of the mean life and decay mode of the 130 keV state are discussed by Fink and Benczer-Koller (8). The half-life of the state has been measured electronically, and the transition matrix element for excitation has been derived from Coulomb excitation data (12). The combination of the Coulomb excitation yield, the internal conversion coefficient (8) $\alpha = 1.76 \pm 0.19$, and the branching ratio (8) $P_{co} = 0.060 \pm 0.008$ for the crossover decay to ground, yields a half-life $\tau_{1/2} = (0.414 \pm 0.014)$ ns in excellent agreement with a recent (15) Mossbauer determination of the line width, $\Gamma = (4.4 \pm 0.4)$ mm/sec, equivalent to $\tau_{1/2} = (0.49 \pm 0.05)$ ns. Wilenzick et al. (15) do not indicate the thickness of the Pt absorber used.

The resonance line from the 130 keV state is a factor of 5 narrower than that from the 99 keV transition and therefore more amenable to isomer shift studies. However, even at 20°K, the 130 keV transition recoilfree fraction, $f = 2.8\%$, is much smaller than that of the 99 keV transition, and the effect observed with a 0.011-inch absorber is only 0.26% (10).

Lattice Dynamics

Two of the more direct techniques used in the study of lattice dynamics of crystals have been the scattering of neutrons and of x-rays from crystals. In addition, the phonon vibrational spectrum can be inferred from careful analysis of measurements of specific heat and elastic constants. In studies of Bragg reflection of x-rays (which involves no loss of energy to the lattice), it was found that temperature has a strong influence on the intensity of the reflected lines. The intensity of the scattered x-rays as a function of temperature can be expressed by $I(T) = I_o\, e^{-2W(T)}$ where $2W(T)$ is called the Debye-Waller factor. Similarly, in the Mossbauer effect, gamma rays are emitted or absorbed without loss of energy and without change in the quantum state of the lattice by a nucleus bound in a crystal. The fraction f of recoilless gamma rays emitted or absorbed by a nucleus in a crystal lattice is e^{-2W} where

$$2W = \frac{2R}{\hbar^2} \int_0^{\omega_m} \left(\frac{\hbar\omega}{2} + \frac{\hbar\omega}{e^{\hbar\omega/kT} - 1} \right) \frac{G(\omega)}{\omega^2} d\omega$$

where R = recoil energy of nucleus and $G(\omega)$ = frequency distribution, normalized so that

$$\int_0^{\omega_m} G(\omega) \, d\omega = 1$$

If the lattice is assumed to have a Debye frequency spectrum with $\omega_m = k\theta_{DW}/\hbar$, then

$$2W = \frac{6R}{k\theta_{DW}} \left[\frac{1}{4} + \left(\frac{T}{\theta_{DW}} \right)^2 \int_0^{\theta_{DW}/T} \left(\frac{x}{e^x - 1} \right) dx \right]$$

where θ_{DW} is the effective Debye temperature. Since the real frequency spectrum, $G(\omega)$, may be considerably different from a harmonic Debye spectrum, the effective Debye temperature appropriate to any other quantity involving a different average over the spectrum may be appreciably different from θ_{DW}. Thus, for example, the Debye temperature θ_c pertinent to the heat capacity will vary much more rapidly with temperature than θ_{DW}. In principle, however, from knowledge of the heat capacity, lattice constants, and elastic constants as a function of temperature, one can learn enough about the frequency spectrum to predict the Debye-Waller factor. Feldman and Horton (7) have made such a calculation in the quasiharmonic approximation in which the lattice potential in Pt metal is assumed to be harmonic, and of all anharmonic effects, only the

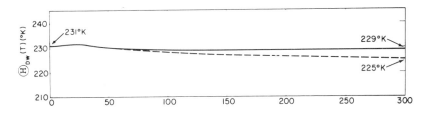

Physical Review

Figure 4. Plot of $\theta_{DW}(T)$ vs. T. The solid line gives the value of $\theta_{DW}(T)$ referred to the 0°K volume. The broken line shows the effect of thermal expansion (7).

thermal expansion is taken into account. Their conclusions are that the Debye temperature should be $\theta_{DW} = (231 \pm 3)\,°K$ at $0°K$ and should have a very slight temperature dependence (Figure 4). Experimentally, from the analysis of the temperature dependence of the absorption spectra obtained with a source of ^{195}Au embedded in a natural platinum matrix and a natural platinum absorber, Harris, Rothberg, and Benczer-Koller (*10*) obtained an extrapolated $\theta_{DW} = (234 \pm 6)\,°K$ at $0°K$ in excellent agreement with theoretical predictions for the platinum cubic lattice. Actually, the quantity that is measured is $f/(1 + \alpha)$, namely a combination of the recoilless fraction f and of the conversion coefficient α. The determination of θ_{DW} is then very sensitive to the choice of α (Figure 5). If the experimental 2W is fitted to the theoretical predictions of Feldman and Horton, then a value for the conversion coefficient α can be obtained, $\alpha = 7.0 \pm 0.3$. This value of the internal conversion coefficient is more

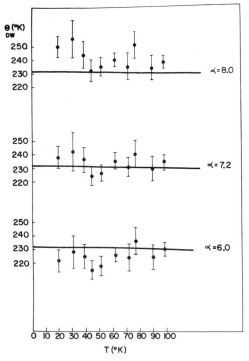

Physical Review

Figure 5. Experimental values (9) of θ_{DW} for various values of the internal conversion coefficient vs. temperature. The solid line was obtained from the analysis of specific heat measurements and other thermodynamic data (7).

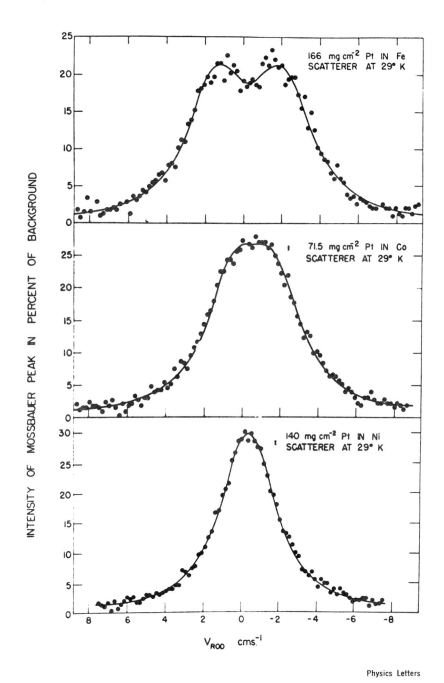

Figure 6. Mossbauer spectra with Pt in ferromagnetic alloys (2)

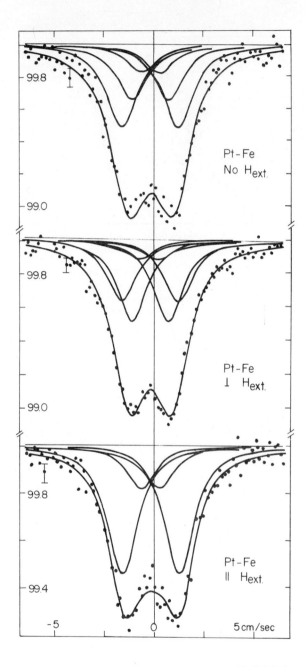

Figure 7. Mossbauer spectra for Pt–Fe absorbers in different field configurations (1)

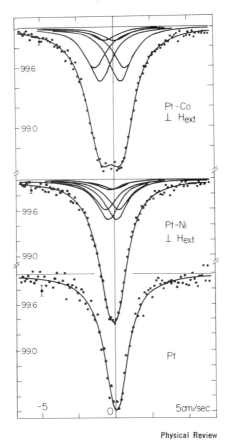

Figure 8. Mossbauer spectra for absorbers of Pt–Co, Pt–Ni, and Pt (1)

precise than that which can be obtained by any of the nuclear spectroscopy methods available to date.

Hyperfine Interaction

The Mossbauer absorption of scattering experiments were carried out by Benczer-Koller, Harris, and Rothberg (3), Atac, Debrunner, and Frauenfelder (2), Agresti, Kankeleit, and Persson (1), and Buyrn, Grodzins, Blum, and Wulff (6) with a single line source and absorbers of Pt–Fe alloys ranging in composition from $Pt_{0.03}Fe_{0.97}$ to $Pt_{0.50}Fe_{0.50}$, Pt–Co, and Pt–Ni alloys as well as Pt_3Fe. A typical spectrum is shown in Figure 6. The expected 6-line pattern corresponding to a $3/2 \to 1/2$ magnetic dipole transition is not resolved, and only two peaks can be observed. From a 6-line fit to the observed pattern, Harris et al. (3) were

Table I. Magnetic Moment of the 99 keV Excited State in

Absorber	$t_a \left(\dfrac{Mg}{Cm^2} \right)$	T_s, °K	T_a, °K
$Pt_{0.1}Fe_{0.9}$	229	20	20
Pt_3Fe	418	20	20
$Pt_{0.3}Fe_{0.7}$	166	29	29
$Pt_{0.07}Co_{0.93}$	71.5	29	29
$Pt_{0.07}Ni_{0.93}$	140	29	29
$Pt_{0.03}Fe_{0.97}$	30–60	4.2	4.2
$Pt_{0.03}Co_{0.97}$	50	4.2	4.2
$Pt_{0.03}Ni_{0.97}$	50	4.2	4.2
$Pt_{0.03}Fe_{0.97}$	52	4.2	4.2
⋮			
$Pt_{0.50}Fe_{0.50}$	521		
$Pt_{0.03}Fe_{0.97}$	Calorimetric determination		
$Pt_{0.1}Fe_{0.9}$	Polarized neutron-spin resonance		
Average			

Table II. Width, Isomer Shift, Magnitude of the ^{195}Au in Various Matrices and

Source Matrix	$t_a \left(\dfrac{Mg}{Cm^2} \right)$ Pt Foil	T_s, °K	T_a, °K	$\Gamma \left(\dfrac{Cm}{Sec} \right)$
Cu	326	4.2	4.2	2.04 ± 0.10
Be	326	4.2	4.2	~ 2.5
Ir	610	4.2	4.2	3.05 ± 0.15
Pt	104	20.0	20.0	2.1 ± 0.2
^{195m}Pt in Pt	326	4.2	4.2	–

only able to obtain a relationship between H_{int}, the effective field at the nucleus, and the ratio μ_e/μ_g of magnetic moments of the excited and ground states. The three other groups altered the absorption or scattering patterns by polarizing the absorbers with an external field applied either parallel or perpendicular to the gamma ray emission (Figures 7 and 8). They thus were able to obtain unique values for the sign and magnitude of μ_e and H_{int}. These results, as well as internal fields obtained from a calorimetric determination (11) and a polarized neutron spin resonance

^{195}Pt and Internal Magnetic Field for Various Pt–Fe Alloys

μ_e	H_{int} MG	Ref.
$-0.9 < \mu < 0.18$	$-1.2 < H < -2.90$	3
-0.645 ± 0.150	-1.24 ± 0.15 -0.77 ± 0.07 -0.34 ± 0.08	2
-0.615 ± 0.045	-1.119 ± 0.04 -0.86 ± 0.03 -0.36 ± 0.04	1
-0.585 ± 0.120	-1.29 ± 0.09	6
	$|H_{int}| = 1.39$	11
	$|H_{int}| = 1.08$	13
-0.615 ± 0.041		1,2,6

Effect, and Effective Debye Temperature for Sources of Natural Platinum Absorber (6)

$\delta \left(\dfrac{Cm}{Sec} \right)$	Effect, %	θ_D, °K	Ref.
<0.05	6.2 ± 0.3	220 ± 20	5
	6.0 ± 0.8	220 ± 30	
<0.05	11.9 ± 1.0	315 ± 35	
	5.2 ± 0.2	234 ± 6	
–	~6		

experiment (13) on similar alloys, are shown in Table I. These latter values of H_{int} do not agree too well with the Mossbauer effect results. H_{int} is smaller for the Pt–Co and Pt–Ni alloys than for the Pt–Fe case.

However, the ratio of H_{int} to the effective magnetic moment of the host atom is almost constant for the three alloys, as is to be expected if conduction electron polarization is the cause of the internal fields. Accurate theoretical estimates of the contributions to H_{int} from conduction electron polarization and from core polarization have not been made yet.

Debye Temperatures for Platinum in Pt, Be, Cu, and Ir

In most experiments reported, ^{195}Au sources embedded in platinum matrices were used. For these sources, the recoilless fraction, even at liquid helium temperatures, is small, and the resonance effects are of the order of a few per cent at best. Buyrn and Grodzins (5) searched for

Table III. Widths, Isomer Shift, and Mossbauer Fractions

Absorber	$t_a\left(\dfrac{Mg}{Cm^2}\right)$	T_s, °K	T_a, °K	$\Gamma\left(\dfrac{Cm}{Sec}\right)$
Pt	217	20	20	
$Pt_{0.1}Fe_{0.9}$	229	20	20	
$Pt_{0.3}Fe_{0.7}$	147	20	20	
$Pt_{0.5}Fe_{0.5}$	216	20	20	
Pt_3Fe	418	20		
Pt	32	29	29	1.91 ± 0.08
	122	29	29	1.98 ± 0.07
	580	29	29	2.14 ± 0.06
	580	92	92	2.46 ± 0.07
$Pt_{0.03}Fe_{0.97}$	166	29	29	2.05
$Pt_{0.07}Co_{0.93}$	71.5	29	29	2.03
$Pt_{0.07}Ni_{0.93}$	140	29	29	2.01
Pt	50	4.2	4.2	1.80 ± 0.04
$Pt_{0.03}Fe_{0.97}$	30–60	4.2	4.2	1.74 ± 0.003
$Pt_{0.03}Co_{0.97}$	50	4.2	4.2	1.86 ± 0.004
$Pt_{0.03}Ni_{0.97}$	50	4.2	4.2	1.78 ± 0.006
PtO				
PtO_2				
$PtCl_2$				
$PtCl_4$				
$Pt_{0.20}Au_{0.80}$				
$Pt_{0.07}Al_{0.93}$				

environments for which a higher Debye temperature and hence a larger effect could be obtained. They used sources of ^{195}Au and ^{195m}Pt in Cu, Be, and Ir matrices (Table II). Ir is by far the best host material, with $\theta_{DW} = (315 \pm 35)$°K, as compared with $\theta_{DW}(Pt) = (234 \pm 6)$°K. In all cases, the Debye-Waller temperature θ_{DW} is only in rough agreement with the predicted values $\theta_{eff} = \theta_{host}(M_{host}/M_{imp})^{1/2}$ for an impurity atom bound with the same force constant as the host atoms.

Isomeric Shift

Agresti, Kankeleit, and Persson (*1*) studied the Mossbauer effect with alloys of 20 atom % Pt in Au and 0.7 atom % Pt in Al, as well as with compounds such as PtO, PtO_2, $PtCl_2$, and $PtCl_4$ (Table III) in order to obtain further information about isomeric shifts and possibly electric

for Various Ferromagnetic Alloys and Platinum Compounds

$\delta\left(\dfrac{Cm}{Sec}\right)$	f_s	f_a	Ref.
1.70 ± 0.20		0.129 ± 0.008	*3,9*
		0.269 ± 0.041	
	0.089 ± 0.002	0.243 ± 0.026	
		0.116 ± 0.008	
		0.096 ± 0.010	
−0.16 ± 0.04			*2*
−0.10 ± 0.02	0.096 ± 0.015	0.105 ± 0.015	
−0.14 ± 0.07			
−0.12 ± 0.14	0.02 ± 0.003	0.02 ± 0.003	
−0.21 ± 0.02			*2*
−0.35 ± 0.03			
−0.23 ± 0.02			
−0.006 ± 0.017			*1*
−0.190 ± 0.011			
−0.196 ± 0.009			
−0.165 ± 0.008			
−0.034 ± 0.011			
−0.040 ± 0.008			
−0.010 ± 0.020			
−0.03 ± 0.03			
+0.075 ± 0.002			
−0.205 ± 0.001			

quadrupolar interactions. They have observed isomeric shifts (Figure 9) which decrease with decreasing electronegativity of the host element. If decreasing the electronegativity of the host material reflects an increase in the electron density at the impurity nucleus, the sign of the isomeric shifts implies that the mean square radius for the ^{195}Pt first excited state is smaller than that of the ground state. This result is in agreement with the prediction of Uher and Sorensen (*14*) for the relative effective radii

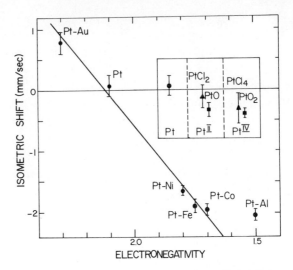

Figure 9. Plot of the isomeric shift of platinum alloys vs. the electronegativity of the host element. The deviation for Pt–Al from the general trend might occur because the platinum is not in a solid solution. In the insert, isomeric shifts for some platinum compounds are shown related to the valency of platinum (1).

of ^{195}Pt nuclei with the valence neutron in the $3p_{3/2}$ or the $3p_{1/2}$ states. Both the 99 and 130 keV states exhibit relatively small collective effects [B(E2) ≈ 10 W.u.] compared with the higher excited states of ^{195}Pt, and therefore the root mean square radii should be controlled mostly by the monopole core polarization contribution. Under these circumstances, the isomer shift for the 130 keV state should be vanishingly small, as the effective radii for the $2f_{5/2}$ and $3p_{1/2}$ neutron states are almost equal.

Acknowledgment

This work was supported by the National Science Foundation.

Literature Cited

(1) Agresti, D., Kankeleit, E., Persson, B., *Phys. Rev.* **1967**, 155, 1339.
(2) Atac, M., Debrunner, P., Frauenfelder, H., *Phys. Letters* **1966**, 21, 699.
(3) Benczer-Koller, N., Harris, J. R., Rothberg, G. M., *Phys. Rev.* **1965**, 140, B547.
(4) Blaugrund, A. E., *Phys. Rev. Letters* **1959**, 3, 226.
(5) Buyrn, A. B., Grodzins, L., *Phys. Letters* **1966**, 21, 389.
(6) Buyrn, A. B., Grodzins, L., Blum, N. A., Wulff, J., *Phys. Rev.* **1967**, 163, 286.
(7) Feldman, J. L., Horton, G. K., *Phys. Rev.* **1965**, 137, A1106.

(8) Fink, T., Benczer-Koller, N., *Nucl. Phys.* **1969**, A138, 337.
(9) Harris, J. R., Benczer-Koller, N., Rothberg, G. M., *Phys. Rev.* **1965**, 137, A1101.
(10) Harris, J. R., Rothberg, G. M., Benczer-Koller, N., *Phys. Rev.* **1965**, 138, B554.
(11) Ho, J. M., Phillips, N. E., *Phys. Rev.* **1965**, 140, A648.
(12) Keszthelyi, L., Cameron, J. A., private communication.
(13) Stolovy, A., *Bull. Am. Phys. Soc.* **1965**, 10, 17.
(14) Uher, R. A., Sorensen, R. A., *Nucl. Phys.* **1966**, 86, 1.
(15) Wilenzick, R. M., Hardy, K. A., Hicks, J. A., Owens, W. R., *Phys. Letters* **1969**, 29A, 670.

RECEIVED December 16, 1969.

11

Double Bond Isomerization as a Product-Controlling Factor in Hydrogenation over Platinum Group Metals

PAUL N. RYLANDER

Engelhard Industries, Newark, N. J. 07105

> *A useful endeavor in catalysis is to describe properties of catalysts in terms that will correlate various phenomena and permit prediction. One such parameter is the relative tendency of catalysts to promote double bond migration during hydrogenation. This tendency relates to such diverse events as catalyst poisoning, changes in the organic function, hydrogenolysis, loss of optical activity, and selectivity in hydrogenation of aliphatic and aromatic systems.*

All of the elements in the platinum group metals make hydrogenation catalysts, and among them they can catalyze the reduction of most organic functional groups. Both the rate and selectivity depend markedly on the metal. The number of organic compounds and the products derivable from them by hydrogenation is very large, and it is of considerable practical value to search for a few characteristics of catalysts that will permit some correlation of the catalytic metal with the type of product obtained. One of these useful characteristics is the relative ability of various catalysts to promote double bond isomerization in olefins prior to their saturation. The concept is applicable not only to olefins *per se* but also to compounds such as acetylenes and aromatics that are converted to olefins during hydrogenation. In the presentation that follows, a number of diverse phenomena will be correlated with this property.

Double Bond Migration

Olefins may undergo a facile double bond migration in the presence of hydrogen and a platinum metal catalyst. A relative order (palladium >> ruthenium > rhodium > platinum >> iridium) established (2) for

1-pentene, for the tendency to promote double bond migration during hydrogenation, is probably general for most olefinic systems. The extent of isomerization also depends importantly on the substrate. For migration to occur, the allylic hydrogen to be removed must be sterically accessible to the catalyst and on the same side of the molecule as the entering hydrogen (5). The mechanism involved may be complex (3). Migration is diminished by the presence of heavy metals, hydroxides, amines, and elevated pressure (15).

Selectivity Among Olefinic Types

A useful generality in predicting selectivity is that the rate of hydrogenation falls, both absolutely and competitively, as steric hindrance around the double bond is increased (23). On this basis in a mixture of olefins, terminal olefins would be expected to be saturated preferentially to internal olefins, if a prior rapid equilibration by isomerization did not ensue.

This thesis was demonstrated (1) in the selective hydrogenation of the pairs of olefins shown in Table I, over ruthenium-on-carbon, a catalyst with relatively low isomerization activity. The experiments were carried out by partial hydrogenation of a mixture of 1 mole of each olefin, and the reaction was interrupted and analyzed after absorption of 1 mole of hydrogen. Those compounds underlined in Table I were reduced with high selectivity in preference to the other member of the pair. This high degree of selectivity was limited to those pairs of olefins

Table I. Competitive Hydrogenation of Olefins by Ruthenium on Carbon

<u>4-Methyl-1-pentene</u> and 2-methyl-2-pentene	Selective
<u>4-Methyl-1-pentene</u> and 2-methyl-1-pentene	Selective
<u>4-Methyl-1-pentene</u> and 2-octene	Selective
<u>4-Methyl-1-pentene</u> and cyclohexene	Selective
2-Methyl-2-pentene and 2-methyl-1-pentene	Not selective
2-Methyl-2-pentene and 2-octene	Not selective
2-Methyl-2-pentene and <u>1-octene</u>	Selective
2-Methyl-2-pentene and <u>cyclohexene</u>	Not selective
2-Methyl-1-pentene and 2-octene	Not selective
2-Methyl-1-pentene and <u>1-octene</u>	Selective
2-Methyl-1-pentene and <u>cyclohexene</u>	Not selective
2-Octene and <u>1-octene</u>	Selective
2-Octene and <u>cyclohexene</u>	Not selective
<u>1-Octene</u> and cyclohexene	Selective

containing one terminal and one internal olefin; other pairs were reduced with a much lower selectivity. This same type of discrimination between internal and terminal olefins also has been demonstrated when the two types of double bonds were in the same molecule. These hydrogenations were carried out with water as a solvent, if indeed solvent is the correct word to use with olefins, but if there were no water present there was no reduction. This is a common characteristic of ruthenium catalysis at low pressure. At high pressure, many solvents can be used, but even here water sometimes has a strikingly beneficial effect on the rate (32).

Selectivity of the type found with ruthenium was not possible when palladium catalysts were used. For instance, hydrogenation of a mixture of 1- and 2-octene was completely nonselective over palladium catalysts. This lack of selectivity resulted from the high isomerization activity of palladium; when the reaction was stopped at only one-tenth of completion, all 1-octene had disappeared by migration of the terminal double bond inward.

Migration to Inaccessible Positions

Trivial reductions at times may be rendered difficult or impossible by improper selection of catalysts. This is demonstrated (13) by the hydrogenation of pulchellin, I. Over platinum oxide in ethanol, pulchellin was reduced to dihydropulchellin, II. On the other hand, reduction of pulchellin over palladium-on-carbon or palladium-on-calcium carbonate afforded dihydropulchellin in 60–80% yield accompanied by the isomerized product, isopulchellin, III. These results are in keeping with the greater tendency of palladium to cause double bond migration. It would be most difficult to convert isopulchellin to dihydropulchellin by catalytic hydrogenation, for any attempt to do so probably would result in hydrogenolysis of the allylic carbon–oxygen bond.

There are many other examples of double bond migration to a tetra-substituted position, and these include hydrogenations even over platinum, a catalyst with a relatively low isomerization activity (*24*). One example, the attempted hydrogenation of the exocyclic double bond of hysterin over platinum oxide, resulted in isomerization to isohysterin, which was resistant to hydrogenation under the conditions of the reaction. Hysterin (*39*) was reduced successfully over ruthenium dioxide at elevated temperatures and pressures. It is expected that the elevated pressures used with ruthenium would diminish isomerization, and the vigorous conditions with ruthenium also might be sufficient to reduce isohysterin if it formed.

Changes in the Organic Function

Isomerization during hydrogenation may alter the functional groups present in the molecule. For instance, reduction of cyclohexen-2-ol over 5% platinum-on-carbon occurred with rapid absorption of 1 mole of hydrogen at substantially constant rate and quantitative formation of cyclohexanol. On the other hand, reduction over palladium ceased abruptly at about two-thirds of a mole, and the product was a mixture of cyclohexanol and cyclohexanone, the latter arising through double bond migration to the enol of cyclohexanone (*29*).

OH	OH	O
(cyclohex-2-en-1-ol)	→ (cyclohexanol) + (cyclohexanone)	
5% Pd/C	67%	33%
5% Pt/C	100%	0%

This type of transformation may occur also by an exchange reaction. Catalytic hydrogenation of 5-cyclodecen-1-ol over platinum oxide in methanol resulted in absorption of 0.97 molar equivalents of hydrogen and formation of the saturated alcohol. However, when 10% palladium-on-carbon in alcohol was used as a catalyst, reduction was incomplete owing to formation of cyclodecanone. The authors (*8*) explained the exceptional ease of this type of reaction by the spatial proximity of the alcohol and olefin groups, favoring an intramolecular hydrogen transfer. Presumably, palladium is very much more effective in this transfer reaction than platinum, but similar transfer by platinum might pass unobserved, as this catalyst would reduce the ketone so formed to the alcohol; palladium is a relatively poor catalyst for hydrogenation of aliphatic ketones (*6*).

Catalyst Inhibition

With sufficient use, all catalysts undergo deactivation. Two broad types of inhibition may be distinguished. In the first, the catalyst becomes inhibited through the introduction of catalyst inhibitors carried into the system by the substrate, solvent, hydrogen, or even formed by dissolution of the reactor. In the second type, deactivation arises through some inhibitor formed in the reaction itself. An example of the second type is provided by the work of Nishimura and Hama (21). These workers found, contrary to what one might expect by analogy to similar situations, that benzyl alcohol was reduced more smoothly to cyclohexylcarbinol over platinum oxide than over rhodium catalysts. The rate of hydrogenation over rhodium decreased markedly during the course of the reaction, accompanied by the formation of increasing amounts of cyclohexanecarboxaldehyde, a strong inhibitor. Reductions over platinum catalyst, on the other hand, proceeded at a more nearly constant rate and with the accumulation of less aldehyde. The inhibiting cyclohexanecarboxaldehyde arose from isomerization of intermediate 1-cyclohexenylcarbinol. Platinum catalysts were less inhibited than rhodium by the aldehyde on two counts. The intermediate unsaturated carbinol was isomerized less over platinum than rhodium, and the aldehyde once formed was hydrogenated about four to five times more rapidly over platinum.

Similar isomerizations have been observed in the hydrogenation of phenylethanol over rhodium, where one of the intermediate products is cyclohexylmethylketone.

This ketone, unlike cyclohexanecarboxaldehyde, is not an inhibitor and can be obtained as an intermediate in about 40% yield from hydrogenation of acetophenone over rhodium; further reduction affords high yields of cyclohexylmethylcarbinol (28).

Hydrogenolysis of Carbon–Oxygen Bonds

Carbon–oxygen bonds are not cleaved readily by catalytic hydrogenolysis unless they are in some way activated by proximity to another function. Benzylic, vinylic, and allylic oxygen bonds undergo a facile hydrogenolysis, but functions removed from the double bond do not unless a prior isomerization occurs. An example of isomerization leading to hydrogenolysis products is found in the hydrogenation of hept-3-yne-1,7-diol over 5% palladium-on-barium sulfate; in addition to the expected diol, a considerable amount of 1-heptanol was formed (9). The formation of 1-heptanol is best accounted for by the assumption of reduction of the acetylene to the corresponding olefin followed by isomerization of the double bond into an allylic position. Platinum, or better rhodium, would be preferred to palladium in examples of this type. Rhodium would seem particularly desirable on two counts. The isomerization activity of rhodium is much lower than that of palladium, and less allylic alcohol would be formed in the hydrogenation. Additionally, rhodium has found considerable use in the reduction of allylic and vinylic systems when hydrogenolysis was to be avoided (27).

$$HOCH_2CH_2C \equiv CCH_2CH_2CH_2OH \rightarrow$$
$$HOCH_2CH_2CH = CHCH_2CH_2CH_2OH \rightarrow$$
$$HOCH_2CH = CHCH_2CH_2CH_2CH_2OH \rightarrow$$
$$CH_3CH = CHCH_2CH_2CH_2CH_2OH \rightarrow$$
$$CH_3CH_2CH_2CH_2CH_2CH_2CH_2OH$$

Hydrogenolysis of Carbon–Carbon Bonds

A striking example of the effect isomerization may have is found in the hydrogenation of car-3-ene (7). Two different products are obtained in nearly quantitative yield, depending on the catalyst used. Over platinum-on-carbon at 1500 psig, *cis*-carane was obtained in 98% yield, whereas over palladium-on-carbon, quantitative yields of 1,1,4-trimethylcycloheptane could be achieved. Hydrogenolysis with formation of the seven-membered rings was believed to be preceeded by isomerization to car-2-ene. If isomerization is a prerequisite for hydrogenolysis, it is to be expected that less of the hydrogenolysis product would be formed over platinum, since platinum is not as active as palladium for double bond migration.

Formation of 1,1,4-trimethylcycloheptane from car-3-ene depends on two properties of palladium, the strong tendency of palladium to promote isomerization and to hydrogenolyze vinylcyclopropane systems. In the bicyclic unsaturated acid, IV, prior isomerization is not necessary to move the double bond into conjugation, yet here, too, platinum and palladium give quite different results. The saturated bicyclic acid is formed over platinum, whereas over palladium a mixture of cyclohexane-carboxylic acid and benzoic acid results through hydrogenolysis and disproportionation (20).

Disproportionation

Disproportionation is a special form of double bond migration in which the double bond is transferred from one molecule to another. Reactions of this type are especially liable to occur over palladium, and for this reason palladium sometimes is best avoided in olefin hydrogenation when the double bond is contained in an incipient aromatic system. Disproportionation activity in the hydrogenation of cyclohexene (and presumably other incipient aromatic systems will follow the same order) decreases with the metal in the order palladium >> platinum > rhodium (16). An example of the complication that can be caused by disproportionation during hydrogenation is found in the attempted reduction of the dienone, V, to the saturated ketone over palladium-on-carbon in ether. The product mixture consisted of about 20% indane, 20% cis-hexahydro-1-indanone, and 45% 4,5,6,7-tetrahydro-1-indanone (14). Loss of the oxygen function would be expected once the compound has become aromatic, for palladium makes the best known catalyst for selective hydrogenation of an aromatic ketone to an aromatic alcohol and for hydrogenolysis of the aromatic alcohol to the aromatic hydrocarbon.

Loss of Optical Activity

Unsaturated optically active compounds having an asymmetric center accessible to a migrating double bond may undergo racemization on hydrogenation (4). An example is the hydrogenation of the optically active olefin (−)-3,7-dimethyl-1-octene (15). Reduction over platinum oxide, a catalyst of low isomerization activity, resulted in only 3% racemization, but over varying amounts of palladium-on-carbon, the product was 43–52% racemized. Racemization, and therefore by inference isomerization, were decreased by pressure, by catalyst inhibitors, and by bases. Reduction at 1500 psig over palladium-on-carbon gave only 23% racemized product, over a Lindlar catalyst (19) at 1 atm gave 16%, and over palladium-on-carbon in the presence of small quantities of potassium hydroxide or pyridine gave 12 and 18% racemization, respectively. Certainly, in this compound, migration of the double bond to afford the 2-octene will result in racemization if the olefin is desorbed before saturation, but other pathways to racemization also may be operative.

Stereochemistry

The extent of double bond migration during reduction of an olefin may influence markedly the stereochemistry of the resulting product. A striking example is provided by the work of Siegel and Smith (35) on hydrogenation of 1,2-dimethylcyclohexene. Over platinum oxide in acetic acid, this olefin afforded 82% cis-1,2-dimethylcyclohexane, whereas over palladium-on-alumina the product contained 73% trans-isomer. The explanation for this unexpectedly large percentage of trans-isomer lies in double bond migration to the 2-position or to the methylene position. The stereochemistry of the product is in a sense the resultant of competition between rates of hydrogenation and rates of isomerization of the individual olefinic species.

In general, it seems easier to rationalize a stereochemical result than it is to predict it. The stereochemistry of hydrogenation of olefins is such that hydrogen usually is added as if by cis addition from the catalyst to the side of the molecule that is adsorbed on it (*34*), but unfortunately it is not always easy to decide which side of the molecule will adsorb on the catalyst. The consequence of double bond migration on the stereochemistry of the product is frequently, therefore, not easy to predict.

Hydrogenation of Aromatic Rings

Aromatic compounds are reduced over the six platinum metal group catalysts at widely different rates, as expected, but additionally the products of reduction frequently vary with the metal used. Many of these results may be correlated in terms of two parameters not obviously connected to aromatic properties: the relative tendencies of these catalysts to promote double bond migration in olefins and to promote hydrogenolysis of vinylic and allylic functions.

Hydrogenolysis of Vinylic and Allylic Functions

A generalization derived from many studies is that the tendency toward hydrogenolysis of vinylic and allylic oxygen, nitrogen, and halogen increases in the order ruthenium $\lesssim$ rhodium $<<$ palladium $\lesssim$ platinum $\lesssim$ iridium. In vinylic systems especially, platinum favors hydrogenolysis more than palladium; in allylic systems, the trend is not so clear but appears to be the reverse (*26*). Iridium is placed in this sequence on the basis of limited data.

Partial Hydrogenation of Aromatic Rings

Although hydrogenation of carbocyclic aromatics usually leads to the fully saturated derivative, there is abundant evidence that many reductions proceed through intermediate olefins. Hydrogenation of cresols was assumed to proceed through formation of all kinds of 1,2-dihydrocresols produced in equal amounts (*38*) and hydrogenation of xylenes to afford equal quantities of all tetrahydroxylenes as intermediates (*36*).

Occasionally with catalysts having high activity for aromatic ring reduction and relatively low activity for olefin saturation, partial reduction to olefins may be useful in synthesis. For example, hydrogenation of benzene over ruthenium afforded cyclohexene (*12*) in 20% yield and cis-Δ^2-tetrahydroterephthalic acid was obtained in 73% yield from terephthalic acid (*31*).

Prior to saturation, intermediate olefins may undergo migration, and it is the extent of this step that accounts in large part for the varying results over different catalysts.

Hydrogenation of Anilines

Hydrogenation of anilines usually affords a saturated amine as the major product, but several side reactions, including hydrogenolysis, reductive hydrolysis, and reductive coupling, may accompany and even dominate the reduction. Hydrogenolysis is important only in certain activated molecules (*10*).

Reductive Hydrolysis. Reductive hydrolysis of anilines to cyclohexanones, assumed to go through an imine-type intermediate, clearly involves isomerization (*18*).

$$\text{PhN(CH}_3)_2 \xrightarrow[\text{H}^+]{2\text{H}_2} \left[\text{C}_6\text{H}_{10}\text{=N(CH}_3)_2 \right] \xrightarrow{\text{H}_2\text{O}} \text{C}_6\text{H}_{10}\text{=O}$$

Palladium, by far the most effective of platinum metals for isomerization, is also the most effective in this reaction. The isomerization involves not only tautomerization of an enamine to an imine, but additionally, isomerization of double bonds remote from the amine into a vinylic position. Platinum, ruthenium, and rhodium afford mixtures of less cyclohexanols and more dimethylcyclohexylamines (*17*).

Under vigorous conditions (100°C and 1000 psig) in 25% aqueous sulfuric acid, the yield of cyclohexanol plus cyclohexanone from dimethylaniline was 90, 75, and 13% over 5% palladium-, 5% rhodium-, and 5% platinum-on-carbon, respectively (*33*). The decreasing yield parallels the decreasing tendency for migration. Reductive hydrolysis is favored by substitution on the nitrogen atom attributed in part to the relative difficulty of hydrogenating hindered enamines.

Reductive Coupling. Formation of dicyclohexylamine (*25*) in hydrogenation of aniline probably involves addition of cyclohexylamine to an imine and subsequent hydrogenolysis of a carbon–nitrogen bond (*11*).

$$\text{PhNH}_2 \rightarrow \left[\text{C}_6\text{H}_{11}\text{NH}_2 \rightleftharpoons \text{C}_6\text{H}_{10}\text{=NH} \right] \rightarrow \text{C}_6\text{H}_{11}\text{NH}_2$$

$$\text{C}_6\text{H}_{10}(\text{NH}_2)\text{-NH-C}_6\text{H}_{11} \rightarrow \text{C}_6\text{H}_{11}\text{-NH-C}_6\text{H}_{11} + \text{NH}_3$$

The per cent of dicyclohexylamine formed in hydrogenation of aniline increases with catalyst in the order ruthenium < rhodium << platinum, an order anticipated from the relative tendency of these metals to promote double bond migration and hydrogenolysis (30). Small amounts of alkali in unsupported rhodium and ruthenium catalysts completely eliminate coupling reactions, presumably through inhibition of hydrogenolysis and/or isomerization. Alkali was without effect on ruthenium or rhodium catalysts supported on carbon, possibly because the alkali is adsorbed on carbon rather than metal (22).

Phenols

On hydrogenation, phenols may undergo saturation to the cyclohexanol, hydrogenolysis of the carbon–oxygen bond, or partial hydrogenation to the cyclohexanone.

Hydrogenolysis

Partial reduction of phenols affords mixtures of allylic and vinylic alcohols. From the generality derived for aliphatic systems, the most hydrogenolysis of this mixture is expected with platinum, palladium, and iridium catalysts, and much less with rhodium and ruthenium, an expectation substantiated in practice. For example, hydrogenation of resorcinol in neutral medium affords 20, 19, and 70% cyclohexanediol over palladium-, platinum-, and rhodium-on-carbon, respectively (29). Many examples attest to the value of rhodium and ruthenium at elevated pressure in avoiding hydrogenolysis.

Partial Reduction to Ketones

Ketones arise from phenols by isomerization of unsaturated alcohols (37). Palladium is the most suited for this type of reaction because of its high isomerization activity coupled with a very low rate of reduction of the resulting ketones (6). Excellent yields of ketones often may be obtained; rhodium will give at times quite substantial yields of cyclohexanones (50–65% methylcyclohexanones from cresols) (38), but in other reductions such as resorcinol, little ketone accumulates over either rhodium or platinum under conditions where it is a major product over palladium (29).

Whether or not ketone accumulates in reduction of phenols depends additionally on the relative rates of hydrogenation of ketone and phenol in competition. For instance, equimolar mixtures of resorcinol and cyclohexanone in ethanol were partially hydrogenated over several catalysts;

over palladium-on-carbon, resorcinol was reduced to the exclusion of cyclohexanone, whereas over platinum-on-carbon the reverse was true. No such marked preference for substrate was observed in reduction over rhodium-on-alumina or rhodium-on-carbon (29).

The problem of ordering platinum group metals for the relative rate of reduction of various functions in competition is, with few exceptions, unsolved. A generalized solution would do much to aid in the proper choice of metal for selective reduction of bifunctional compounds.

Literature Cited

(1) Berkowitz, L. M., Rylander, P. N., *J. Org. Chem.* **1959**, 24, 708.
(2) Bond, G. C., Rank, J. S., *Proc. Intern. Congr. Catalysis, 3rd*, W. M. H. Sachtler *et al.*, Eds., **1965**, 2, 1225.
(3) Bond, G. C., Wells, P. B., *Advan. Catalysis* **1964**, 15, 91.
(4) Bonner, W. A., Stehr, C. E., doAmaral, J. R., *J. Am. Chem. Soc.* **1958**, 80, 4732.
(5) Bream, J. B., Eaton, D. C., Henbest, H. B., *J. Chem. Soc.* **1957**, 1974.
(6) Breitner, E., Roginski, E., Rylander, P. N., *J. Org. Chem.* **1959**, 24, 1855.
(7) Cocker, W., Shannon, P. V. R., Staniland, P. A., *J. Chem. Soc.* **1966**, 41.
(8) Cope, A. G., Cotter, R. J., Roller, G. G., *J. Am. Chem. Soc.* **1955**, 77, 3594.
(9) Crombie, L., Jacklin, A. G., *J. Chem. Soc.* **1957**, 1622.
(10) Freifelder, M., Stone, G. R., *J. Org. Chem.* **1962**, 27, 3568.
(11) Greenfield, H., *J. Org. Chem.* **1964**, 29, 3082.
(12) Hartog, F., U. S. Patent **3,391,206** (July 2, 1968).
(13) Hetz, W., Ueda, K., Inayama, S., *Tetrahedron* **1963**, 19, 483.
(14) House, H. O., Rasmusson, G. H., *J. Org. Chem.* **1963**, 28, 27.
(15) Huntsman, W. D., Madison, N. L., Schlesinger, S. I., *J. Catalysis* **1963**, 2, 498.
(16) Hussey, A. S., Schenach, T. A., Baker, R. H., *J. Org. Chem.* **1968**, 33, 3258.
(17) Kuhn, R., Driesen, H. E., Haas, H. J., *Ann. Chem.* **1968**, 718, 78.
(18) Kuhn, R., Haas, H. J., *Ann. Chem.* **1958**, 611, 57.
(19) Lindlar, H., *Helv. Chim. Acta* **1952**, 35, 446.
(20) Meinwald, J., Labana, S. S., Chadha, M. S., *J. Am. Chem. Soc.* **1963**, 85, 582.
(21) Nishimura, S., Hama, M., *Bull. Chem. Soc. Japan* **1966**, 39, 2467.
(22) Nishimura, S., Shu, T., Hara, T., Takagi, Y., *Bull. Chem. Soc. Japan* **1966**, 39, 329.
(23) Rylander, P. N., "Catalytic Hydrogenation over Platinum Metals," p. 91. Academic Press, New York, 1967.
(24) *Ibid.*, p. 99.
(25) *Ibid.*, p. 355.
(26) *Ibid.*, p. 445.
(27) *Ibid.*, p. 447.
(28) Rylander, P. N., Hasbrouck, L., *Engelhard Ind. Tech. Bull.* **1968**, 8, 148.
(29) Rylander, P. N., Himelstein, N., *Engelhard Ind. Tech. Bull.* **1964**, 5, 43.
(30) Rylander, P. N., Kilroy, M., Coven, V., *Engelhard Ind. Tech. Bull.* **1965**, 6, 11.
(31) Rylander, P. N., Rakonza, N. F., U. S. Patent **3,163,679** (December 22, 1964).
(32) Rylander, P. N., Rakoncza, N., Steele, D. R., Bollinger, M., *Engelhard Ind. Tech. Bull.* **1963**, 4, 95.

(33) Rylander, P. N., Steele, D. R., Engelhard Ind., unpublished observations, 1963.
(34) Siegel, S., *Advan. Catalysis* **1966**, 16, 123.
(35) Siegel, S., Smith, G. V., *J. Am. Chem. Soc.* **1960**, 82, 6082, 6087.
(36) Siegel, S., Smith, G. V., Dmuchovsky, B., Dubbel, D., Halpren, W., *J. Am. Chem. Soc.* **1962**, 84, 3136.
(37) Smith, H. A., Stump, B. L., *J. Am. Chem. Soc.* **1961**, 83, 2739.
(38) Takagi, Y., Nishimura, S., Taya, K., Hirota, K., *J. Catalysis* **1967**, 8, 100.
(39) Vivar, A. R. de, Bratoeff, E. A., Rios, T., *J. Org. Chem.* **1966**, 31, 673.

RECEIVED December 16, 1969.

INDEX

A

Absorption spectra . . 83–4, 86, 88, 91, 140
Accuracy in bond lengths 121
Acetonitrile 111
Activation parameters 117
Adducts
   CO 124
   CS_2 123
   molecular 121
   N_2 123
   nitrosyl 124
   SO_2 123
   tetracyanoethylene 124
Alloys, ordered 7
Allylic functions, hydrogenolysis of 158
Ambidentate ligand 111
Anilines, hydrogenation of 159
Anisotropy, magnetic 11
Antiferromagnetism 1
Aqueous solutions 110
Aromatic rings, hydrogenation of . . 158
Aryl, silyl, and germyl derivatives 104

B

Binuclear nitrides 56
Bonding problems 99
Bond lengths 124

C

Canted-ferrimagnetic structure . . . 9
$CaPd_3O_4$ 34
Catalyst
   hydrogenation 150
   inhibition 154
Cationic complexes 57
$CdPd_3O_4$ 34
Chalcogenides, platinum metal . . . 17
Chemical shift 99
$(CH_3)_3PtClO_4$ 110
CO adducts 124
Coercive force 10
Competitive hydrogenation 151
Conduction bands 18
Conversion coefficient 140
Coupling constant 103
Crystal
   chemistry and synthesis 28
   spectra 95
   structure 7, 18, 44, 120
Crystallite distribution 11
Crystallographic relationships 10

Curie point
   behavior 8
   ferromagnetic 13

D

Debye temperature 145–6
Debye–Waller factor 138
Delocalized π-systems involving Pt
   metals, structures of 129
Dimethylsulfoxide 111
Disordered structure 41
Disproportionation 156
Distortion 132
2,5-Dithiahexane complexes of
   $RhCl_3$ 109
Double bond
   isomerization 150
   migration 150
Dynamics, lattice 138
Dynamic systems 112

E

Electrical
   conductors 17
   resistivity 6
Electron capture decay 136
Electronic spectra 61, 76

F

Fermi contact term 103
Ferromagnetic
   alloys 141
   Curie points 13
Ferromagnetism 1
Fluorophenyl derivatives 104
Frozen-in flux 15

G

Geometrical isomers 109, 112

H

Heteronuclear double resonance
   spectra 105
Hexahalo-complexes, spectra of . . 75
Hydride resonance 117
Hydrido complexes 66, 106
   organo and 112
   preparative methods for 67
   properties of 68
   reaction of 70

Hydrogenation
- anilines 159
- aromatic rings 158
- catalysts 150
- competitive 151

Hydrogenolysis 155
- of vinylic and allylic functions .. 158

Hyperfine interaction 143
Hysteresis loops 14

I

Infrared spectra 42, 47, 50
Inhibition, catalyst 154
Interaction, hyperfine 143
Interatomic distances 36, 128
Internal magnetic field 145
Iridium 5, 19, 54, 70, 80, 95, 122
Iron moment 4
Isomeric shift 144, 147
Isomerization, double bond 150

K

Ketones, partial reduction to 160

L

Lattice dynamics 138
Ligand substitution 112
Local moments 4

M

Magnetic
- anisotropy 11
- circular dichroism 84
- moment 4, 144
- ordering 17
- permanent 1
- properties 1
- susceptibility 1

Magnetization 6, 8
- saturation 10

Marcasite structure 20
Mass susceptibility 2
Metallic properties 18
Metal–olefin bond 113
Mg_2PdO_4 37
Migration 152
- double bond 150

Molecular adducts 121
Moment
- iron 4
- local 4
- magnetic 4

$MO-PtO_2$ system 49
MO_2-PtO_2 system 43
$M_2O_3-PtO_2$ system 47

Mossbauer
- fractions 146
- spectra 142–3
- spectroscopy 135

N

Neutrons, scattering of 138
Nitrosyl adducts 124
N_2 adducts 123
Nuclear properties 99
NMR spectra 98
Nitride ion 54
Nitrido complexes 54

O

Olefins 150
Optical activity, loss of 157
Ordered alloys 7
Ordering alloy 10
Organic function, changes in 153
Organo and hydrido derivatives .. 112
Osmium 14, 54, 80, 95, 98

P

Palladium .. 1, 21, 28, 129, 152, 156, 159
- oxides 28

Paramagnetism 99
PbO_2 46
$PbPdO_2$ 29
Phenols 160
Platinum 3, 9, 21, 39, 76, 81,
                          86, 98, 125, 135, 154
- alkyls 103
- dioxide, reaction of 39
- metal chalcogenides 17
- (II) phosphine complexes 102

Predicted proton resonance
- spectrum 107

Preparative methods for hydrido
- complexes 67

Properties of hydrido complexes .. 68
^{195}Pt chemical shifts 99
β-PtO_2 43
Pyrite structure 18
Pyrochlore
- series 47
- structures 48

R

Racemization 157
Reactions of hydrido complexes .. 70
Recoilless resonance experiments .. 135
Remanence 15
Resonance effect 137
Rhodium 4, 7, 19, 24, 122, 154
Rotation 113
Ruthenium 14, 17, 54, 123, 151, 158
Rutile structure 46

S

Sandwich-like structure 21
Saturation magnetization 10
Scattering of neutrons and x-rays .. 138
Selectivity 151
Signs of the spin–spin coupling
- constants 105

SO$_2$ and CS$_2$ adducts 123
Solvation
   and solvent effects 109
   number 110
Spectra
   absorption 83–4, 86, 88, 91
   crystal 95
   electronic 61
   electronic absorption 76
   infrared 42, 47, 50
   in organic media 89
   Mossbauer 142–3
   of hexahalo-complexes 75
   of PtCl$_4^{2-}$ 81
   predicted proton resonance 107
   vibrational 59, 60, 64
Spectroscopic qualities, ultraviolet 74
Spectroscopy, Mossbauer 135
Spinel 49
Spin–spin coupling 102, 115
Square planar
   complexes 108
   hydrides 101
SrPd$_3$O$_4$ 34
Sr$_2$PdO$_3$ 31
Stereochemistry 105, 124, 157
Structure
   marcasite 20
   pyrite 18
   sandwich-like 21
   tetragonal 23
Susceptibility
   magnetic 1
   mass 2

Symmetry 87
   patterns 46
Synthesis and crystal chemistry .. 28

T

Terminal nitrido complexes 55
Tetracyanoethylene adducts 124
Tetragonal structure 23
Thermal expansion 139
Trans influence 103, 115, 124
Trimethylplatinum (IV)
   complexes 106
   oxinate 108
Trinuclear complex
   iridium 63
   osmium 62

U

Ultraviolet spectroscopic qualities 74
Vibrational spectra 59, 60, 64
Vinylic functions, hydrogenolysis of 158

W

Water exchange 112

X

X-ray diffraction powder data
                          30, 32, 35, 41, 52
X-rays, scattering of 138
X-ray structure determinations ... 120